A SCIENTIFIC BOTTLENECK

How did giraffes get their long necks?

As all students of Darwinian theory know, the giraffe's neck was developed through millions of years of evolution, as giraffes with the longest necks were able to eat foilage high on trees and survived periods of famine while those with short necks died.

This explanation is superbly convincing—except for one detail.

There is increasing unimpeachable evidence that it simply is not true. . . .

THE NECK OF THE GIRAFFE
Darwin, Evolution, and the New Biology

"Lucid and delightful."—THE NEW YORKER

"The only account I know that conveys to the non-specialist reader crucial arguments as capable of affecting our understanding of ourselves as Darwin's account changed the entire late Victorian world."—BOSTON GLOBE

FRANCIS HITCH[...] [...]e Mysterious World: An Atlas of the Unexplained. He is a member of the Royal Archaeological Institute, the Prehistoric Society, and the Society for Physical Research. He lives near London.

The Study of Man from MENTOR and SIGNET Books

THE NECK
OF
THE GIRAFFE

DARWIN, EVOLUTION, AND
THE NEW BIOLOGY

by
Francis Hitching

A MENTOR BOOK
NEW AMERICAN LIBRARY
TIMES MIRROR
New York and Scarborough, Ontario

Copyright © 1982 by Francis Hitching

Published by arrangement with Ticknor & Fields. The hardcover edition was published in the United States by Ticknor & Fields, and in Great Britain by Pan Books Ltd.

MENTOR TRADEMARK REG. U.S. PAT. OFF. AND FOREIGN COUNTRIES
REGISTERED TRADEMARK—MARCA REGISTRADA
HECHO EN CHICAGO, U.S.A.

SIGNET, SIGNET CLASSIC, MENTOR, PLUME, MERIDIAN and NAL BOOKS
are published in the United States by The New American Library, Inc.,
1633 Broadway, New York, New York 10019, in Canada by
The New American Library of Canada Limited,
81 Mack Avenue, Scarborough, Ontario M1L 1M8

Library of Congress Catalog Card Number: 83-60245

First Mentor Printing, July, 1983

1 2 3 4 5 6 7 8 9

PRINTED IN THE UNITED STATES OF AMERICA

Contents

ACKNOWLEDGEMENTS   vii
INTRODUCTION   ix

PART ONE
Impasse

1. The Missing Fossils   3
2. Natural Limits   31
3. Odds Against   45
4. Biological Oddities   66

PART TWO
Alternatives

5. Creation *vs.* Evolution   91
6. Catastrophes and Extinctions   115
7. Patterns of Life   146

PART THREE
Dogma

8. Monkey Business   171
9. Darwin's Legacy   194

REFERENCES   227
BOOKS CITED IN TEXT   235
INTRODUCTORY READING   239
INDEX   243

The Author would like to thank the following for permission to use illustrations appearing in this book: Peter Clarke, Neil Hyslop, Frank V. Lane, the Mary Evans Picture Library, the Trustees of the British Museum (Natural History), the Society for Inter-disciplinary Studies, the *Illustrated London News*, and *Scientific American*. The Trees of Life appearing on pp. 25–28 are taken from a) E. Haeckel, *Natural History of Man*, b) D. Davison, *Our Prehistoric Ancestors*, c) G. de Beer, *Atlas of Evolution*, and d) C. Patterson, *Evolution*. Illustrations of horse ancestry, pp. 17–18, are taken from Purnell's *Encyclopedia of Animal Life*, A. S. Lull, *Organic Evolution*, and O. Abel, *Palaebiologie und Stammesgeschicte*, Jena, 1929, pp. 285–86.

Acknowledgements

I could not have written this book without enjoying the endless patience of so many scientists, in Europe, the United Kingdom, and the United States, who have guided me through the contemporary evolutionary debate. My gratitude goes to them all, and in expressing my thanks to a few of them by name, I hasten to dissociate them from any of the mistaken opinions or errors of fact that, in spite of their help, may still have crept into the text, for these are surely mine. At the British Museum of Natural History, Dr. Peter Forey and Dr. Colin Patterson; at Queen Elizabeth College, Dr. Jeffrey Pollard and Dr. Peter Saunders; at the Open University, Dr. Mae-Wan Ho; Dr. Ted Steele; at Berkeley, Professor Erich Jantsch; at the Institute for Creation Research, Dr. Duane T. Gish; at Swansea, Professor Derek V. Ager; at Harvard, Professor Stephen Jay Gould; at the University of Sussex, Dr. Brian Goodwin.

Many of these have kindly given me permission to quote from their work, as have (in alphabetical order) David Attenborough, Marjorie Grene, Gertrude Himmelfarb, Alice B. Kehoe, Henry M. Morris and Kelly L. Segraves. The following copyright holders are also gratefully acknowledged: the Trustees of the British Museum (Natural History), the Evolution Protest Movement, Family Services of Zurich (for the leaflet on Professor Poopfossil), Massachusetts Institute of Technology, Pergamon Press, Presbyterian and Reformed Publishing Company, Rainbird Publishing Group, Sianauer Associates.

Finally I would like to acknowledge my debt to Arthur Koestler, whose inspired books *The Ghost in the Machine* and *Janus* anticipated many of the arguments that are currently raging in evolutionary biology, and gave me some years ago my first nagging feeling that what Darwin wrote wasn't the end of the story.

Introduction

In April 1882, Charles Darwin died peacefully of a heart attack at his family home in Kent. His great theory, the basis of all modern biology teaching, had come to be accepted with a fervor close to reverence. In the annals of science, his is a name which ranks with the likes of Copernicus, Galileo, Newton and Einstein.

Yet as 1982 approached, and with it the centenary of his passing, change was in the wind. Feuds concerning the theory of evolution exploded rancorously in otherwise staid and decorous scientific journals. Entrenched positions, for and against, were established in high places, and insults lobbed like mortar bombs from either side. Meanwhile the doctrine of Divine creation, assumed by most scientists to have been relegated long ago to the pulpits of obscure fundamentalist sects, swept back into the classrooms of American schools.

Darwinism is under assault on many fronts. It seems an appropriate moment for an outsider to inquire how, truly, *The Origin of Species* has passed the test of time.

Given the importance of the issues involved, and the vehemence with which the debate is being fought, perhaps no one can be strictly objective; so far as possible, I have tried to be. This book is written from the standpoint of a neutralist passing through the battlefield, attempting to discern where victory will lie.

The mystery of how life began, and then proliferated into the myriad forms on Earth today, is one that crosses the barriers of science, religion and philosophy. For many years now, as we shall see, it has been taken for granted that Darwin had solved the question once and for all. But had he? When I began to look beyond the orthodox textbook explanations, I found many scientists with private doubts; rather fewer who had come out with them in public; and a handful who went so far as to say that Darwinian evolutionary theory had turned out not to be scientific theory at all.

Thus you will find Darwinism being dismissed as empty rhetoric; and the status of biology likened to the dark ages of pre-Newtonian physics. On the other hand you will come across one of the world's most eminent evolutionary biologists saying that attacks against Darwin's theory are invariably either based on ignorance or politically motivated.

It is an academic row of far-reaching (and frequently entertaining) proportions—potentially one of those times in science when, quite suddenly, a long-held idea is overthrown by the weight of contrary evidence and a new one takes its place. It happened when Darwinism itself initially triumphed; and in this century when geologists, after years of saying that continental land masses were immovable, suddenly changed their minds and agreed that Africa and South America had indeed been joined together at one time, and had gradually drifted apart.

Darwin's theory of how evolution had happened may now be equally vulnerable, and a concept even more profound waiting to come on stage.

—Francis Hitching
London, July 1981

PART ONE

Impasse

"Well, now, if you really understand an argument you will be able to indicate to me not only the points in favour of the argument but also the most telling points against it."

"I suppose so, sir."

"Good. Please tell me, then, some of the evidence against the theory of Evolution."

"Against what, sir?"

"The theory of Evolution."

"But there isn't any sir."

—Master-pupil dialogue quoted by
Professor G. A. Kerkut,
of the University of Southampton,
in *The Implications of Evolution*

CHAPTER ONE

The Missing Fossils

The number of intermediate varieties, which have formerly existed on the earth, must be truly enormous. Why then is not every geological formation and every stratum full of such intermediate links? Geology assuredly does not reveal any such finely graduated organic chain; and this, perhaps, is the most obvious and gravest objection which can be urged against my theory.

—Charles Darwin,
The Origin of Species, 1859

Stroll around a zoo and look at the giraffes and elephants; wander through a natural history museum and gaze at the reconstructed dinosaurs; pick up a piece of limestone and crack it open to reveal the ancient fossil shell inside; swim in the sea and enjoy, even as you avoid it, the translucent membrane of a floating jellyfish, unchanged in form for 550 million years.

Evolution is about how these wonders happened. How do we come to have lions and jellyfish and things? What is it that has brought about life on Earth in all its astounding diversity?

More than a century ago, Charles Darwin believed he had the answer. It was a beguilingly simple one. If you went back through the fossil record, he said, you would find that today's perfectly adapted life forms had emerged as victors over previous less perfect forms. The predecessors of giraffes or elephants were slightly different animals, perhaps less strong or less large. At any rate, they were feebler compared with those around them, and they had been wiped out by what he called natural selection.

Throughout nature, and throughout Earth's history, he saw that creatures and plants living on its surface had been involved in a constant struggle for survival. In each generation, only the fittest made it. These fit varieties were marginally more successful than their parents—they had better eyesight, maybe, or longer legs to let them run faster, or leaves that enabled them to withstand a sudden extreme of cold. Gradually, almost imperceptibly, these novel features accumulated until a quite new sort of animal or plant emerged—a life form unrecognizable from its distant ancestors millions of years before.

This, said Darwin, solved the greatest puzzle of evolution: the origin of new species (a species is usually defined as a living thing that can only reproduce with its own kind). "From the war of nature, from famine and death, the production of higher animals directly follows."

The idea seemed so blindingly obvious, and so satisfyingly complete, that it quickly replaced the Biblical account of creation, and became a new way of looking at the living world. With a few hiccups, it had held its place ever since. Darwin's friend Thomas Henry Huxley, who trounced Bishop Wilberforce in a historic debate on the subject at the British Association in Oxford in 1860, is said to have remarked after reading *The Origin of Species*, "How stupid of me not to have thought of that."

Yet for all its acceptance in the scientific world as the great unifying principle of biology, Darwinism, after a century and a quarter, is in a surprising amount of trouble.

Evolution and Darwinism are often taken to mean the same thing. But they don't. Evolution of life over a very long period of time is a fact, if we are to believe evidence gathered during the last two centuries from geology, paleontology (the study of fossils), molecular biology and many other scientific disciplines. Despite the many believers in Divine creation who dispute this (including about half the adult population of the United States, according to some opinion polls), the probability that evolution has occurred approaches certainty in scientific terms.

We can be as sure about this as we are sure that ancient civilizations once existed on Earth but no longer function. The archaeological record tells us about these relatively recent times, and the fossil record about earlier ones. If you walk along the trails leading down to the depths of a great fissure such as the Grand Canyon, you can see some of the stages of evolution illustrated by the fossils in front of your eyes. The Earth is old, belongs to an even older universe, and life forms have been upon it for about three quarters of its existence.

On the other hand Darwinism (or neo-Darwinism, its modern version) is a theory that seeks to explain evolution. It has not, contrary to general belief, and despite very great efforts, been proved.

This is not so surprising as it may at first seem. The process of

ERA	SYSTEM AND PERIOD	SERIES AND EPOCH	DISTINCTIVE FEATURES	YEARS AGO
CENOZOIC	QUATERNARY	RECENT	Modern man	11 thousand
		PLEISTOCENE	Early man. Northern glaciation	½ to 3 million
	TERTIARY	PLIOCENE	Large carnivores	13 million
		MIOCENE	First abundant grazing mammals	25 million
		OLIGOCENE	Large running mammals	36 million
		EOCENE	Many modern types of mammals	58 million
		PALEOCENE	First placental mammals	63 million
MESOZOIC	CRETACEOUS		First flowering plants, climax of dinosaurs and ammonites followed by extinction	135 million
	JURASSIC		First birds, first mammals dinsaurs and ammonites abundant	
				180 million
	TRIASSIC		First dinosaurs abundant cycads and conifers	
				230 million
PALAEOZOIC	PERMIAN		Extinction of many kinds of marine animals including trilobites Glaciation at low altitudes	
				280 million
	UPPER CARBONIEEROUS		Great coal forests, conifers First reptiles	
				310 million
	LOWER CARBONIFEROUS		Sharks and amphibians ambundant Large and numerous scale trees and seed ferns	
				345 million
	DEVONIAN		First amphibians and ammonites fishes abundant	
	SILURIAN		First terrestrial plants and animals	405 million
				425 million
	ORDOVICIAN		First fishes, invertebrates dominant	
				500 million
	CAMBRIAN		First abundant record of marine life trilobites dominant, followed by extinction of about two-thirds of trilobite families	
				600 million
	PRE-CAMBRIAN		Fossils extremely rare, consisting of primitive aquatic plants. Evidence of glaciation. Oldest dated algae, over 2,600 million years	

evolution is so slow (some 580—600 million years from the earliest sea creatures until now, according to the latest estimate) that we can scarcely hope to see it happening within our own lifetime, except at a trivial level. We can breed dogs or horses or vegetables and see that great variations of size and appearance can be achieved. In the wild, we can watch dark-winged moths gradually replace light-winged moths on soot-blackened tree trunks, because their camouflage is better and birds don't eat so many of them. In the laboratory, we can bombard fruit flies with x-rays and produce mutants with two sets of wings, deformed bodies, and other unpleasantnesses. Darwin's theory is in fact quite good at explaining minor changes of this sort.

But after that, we can do no more than Darwin did, and extrapolate: that is, to infer that over millions of years, these variations and mutations "must have" gradually added up to the evolution of giraffes and jellyfish and things.

The Fossil Record

But is this what really happened? At first glance, it seems as if it shouldn't be too difficult to demonstrate one way or the other. You simply turn to the fossil record and look for evidence of how life evolved over vast stretches of time. Here, just as in archaeology, you can examine the footprints of ancient history, and piece together a jigsaw picture of strange, long-forgotten eras during nature's march towards the present.

Darwin, like every other evolutionist before and since, turned repeatedly to the fossils to give him support for his theory. Naturally, he hoped to find his "finely graduated organic chain" showing the many transitional forms that led up to the giraffes, jellyfish, etc., that we see today.

Anyone who has puzzled about how fossilized seashells come to be on the tops of mountains, or piled hundreds of feet high to form such landmarks as the white cliffs of Dover, must realize how difficult it has been for geologists to place the fossil record in some kind of rational order. Because of the upheavals brought about by continental drift, huge changes have taken place in Earth's landscape, and the rock strata are convoluted in the extreme.

Nevertheless, geologists feel with some confidence that their modern maps are accurate. Even though there is nowhere in the world you can find an unbroken succession of fossils and rocks from the beginning of time until now, it has been possible to correlate fossils from different places and arrive at a fairly consistent history of Earth's evolution. As we shall see later, a new approach to comparative anatomy and the relationship of fossils to each other, developed during the 1970s, has thrown considerable doubt on important aspects of this history. The familiar textbook trees of life may be altogether too complacently drawn.

However, almost everyone still agrees with the broad picture that emerges from the fossil record. Life has evolved, with occasional interruptions, in ever-growing variety and complexity. There have been a number of fairly well-defined periods when certain life forms have dominated the Earth, replaced eventually by more modern forms. And the main evolutionary developments in the animal world have been successively:

1. Minute, single-celled organisms (bacteria and slime).
2. Multi-celled invertebrates (e.g., sponges, sea-food, jellyfish).
3. Fishes with backbones.
4. Amphibians living partly on land.
5. Reptiles (including dinosaurs).
6. Mammals and birds.

The fossil record readily shows this succession of forms, and Darwin hoped it would do more. He believed, and it has remained the conventional wisdom, that these stages were linked. Fish turned into amphibians, amphibians turned into reptiles, reptiles turned into birds, and so on.

Moreover, Darwin was convinced that this process happened bit by bit—minute "improvements" in successive generations gradually led to the emergence of new species. He took this stance in spite of warnings from his friends—for instance, Thomas Huxley, who wrote him a letter on the day *The Origin of Species* was published. Darwin was taking a risk, he said, in becoming inextricably married to the idea of slow and constant evolution as summed up in the Latin tag *Natura non facit saltum* (Nature does not make leaps). "You have loaded yourself with an unnecessary difficulty," he cautioned.

But Darwin was adamant, and Huxley's warning notwithstanding, most museums and textbooks today accept gradualism as readily as they accept natural selection. Indeed, the two are inseparable. The touchstone of neo-Darwinian theory is that evolution results from the natural selection of small, accidental, cumulative changes.

Logically, then, the fossil record ought to show this stately progression. If we find fossils at all, and if Darwin's theory was right, we can predict what the rocks should contain: finely

PANEL 1
Where are the Pre-Cambrian Fossils?

Although a few fairly complex fossils have been discovered in the upper levels of the Pre-Cambrian, their significant absence in general is a well-established phenomenon, accepted by geologists. Most of the explanations ring changes on the idea that the environment in Pre-Cambrian times was unsuitable for the formation and preservation of fossils.

1. *Fossils were destroyed by changes in the rock structure—e.g., melting or compression.*

Without doubt, much metamorphosis (the geological term for substantial changes in rock structure) took place. Marble is a Pre-Cambrian rock that has been formed through limestone crystallizing under great pressure, so that it cannot contain fossils. But at least ten percent of Pre-Cambrian rocks are sedimentary, many of them continuing in an unbroken procession through to Cambrian times.

2. *There was little or no calcium in Pre-Cambrian seawater; therefore the creatures were soft-shelled, and their fossils were not preserved.*

There are a number of objections to this theory. It ignores areas of the Earth where the rocks continue undisturbed (see above). Shells also become hard and strong through the action of other chemicals besides calcium—silicon and chitin particularly. So for the theory to work, you have to assume that all three chemicals were missing from the Pre-Cambrian, and arrived simultaneously by chance in the Cambrian—an improbably tall order.

3. *Pre-Cambrian organisms lived on the seashore, a volatile environment unlikely to preserve fossils.*

But even if the seashore explanation is true (and it is entirely hypothetical, for there is no evidence one way or the other) some fossils would still have been formed and preserved. And it is generally conceded that the search has been so thorough that, if they exist, they ought to have come to light by now.

4. *"Sudden" and "explosive" are relative terms; ten to twenty million years is more than enough time for life to proliferate on an empty globe, once conditions were right.*

This is the most persuasive, as well as the most recent, of all the explanations. There is a general law of growth which predicts a period of steep acceleration in what is called a sigmoidal (s-shaped) curve.

Towards the end, when all the ecological niches are filled, the curve flattens off. Something of this sort can be inferred in early Cambrian times, with the oceans at first containing plenty of space, abundant food, and no competition. New life forms, triggered off by a perhaps quite minor but significant environmental change, rapidly multiplied to take advantage of the situation. Once the point had been reached where the oceans were full of life, evolution slowed down.

But although this may be a plausible outline for what happened in the Cambrian, it still leaves unanswered the crucial question of what sudden event caused the single-celled creatures to develop into highly complex multi-cellular ones; and what evolutionary or biological mechanism there was to permit this to happen. In a sense, it just pushed the problem back earlier in time. The origin of multi-cellularity remains "the enigma of palaeontological enigmas."

graduated fossils leading from one group of creatures to another group of creatures at a higher level of complexity. The minor "improvements" in successive generations should be as readily preserved as the species themselves.

But this isn't so. In fact the opposite is the case, as Darwin himself complained: "Innumerable transitional forms must have existed but why do we not find them embedded in countless numbers in the crust of the earth?"

The Fossil Gaps

His answer at the time was the "extreme imperfection" of the fossil record, and he consoled himself with the thought that as the geological search continued, enough transitional fossils between one species and another would be found to confirm his theory and allay his doubts.

About the imperfection he was right. There are about 250,000 different species of fossil plants and animals in the world's museums. This compares with about 1.5 million species known to be alive on Earth today. Given the known rates of evolutionary turnover, it has been estimated that at least 100 times more fossil species have lived than have been discovered.[1] Clearly,

Earth has seen a lot of life forms we don't know about, and some of them may have been transitional.

But the curious thing is that there is a consistency about the fossil gaps: *the fossils go missing in all the important places*.

When you look for links between major groups of animals, they simply aren't there; at least, not in enough numbers to put their status beyond doubt. Either they don't exist at all, or they are so rare that endless argument goes on about whether a particular fossil is, or isn't, or might be, transitional between this group and that.

Yet there are lengthy periods of history when there is every reason to expect plenty of intermediates. At such times, geological strata straddling an evolutionary change hold an abundance of evidence—the fossils are of good quality, and their timespan on Earth is known with a high degree of accuracy.

Museums have, for instance, countless piles of fossils of the early invertebrate sea creatures, and an equally large number of ancient fishes. Between the two, covering a period of about 100 million years, there ought to be cabinets full of intermediates—indeed, one would expect the fossils to blend so gently into one another that it would be difficult to tell where the invertebrates ended and vertebrates began.

But this isn't the case. Instead, groups of well-defined, easily classifiable fish jump into the fossil record seemingly from nowhere: mysteriously, suddenly, full-formed, and in a most un-Darwinian way. And before them are maddening, illogical gaps where their ancestors should be.

"Probably most people assume that fossils provide a very important part of the general argument that is made in favor of Darwinian interpretations of the history of life. Unfortunately, this is not strictly true," according to David M. Raup, curator of one of the world's finest natural history museums, the Field Museum in Chicago.

> Instead of finding the gradual unfolding of life, what geologists of Darwin's time and geologists of the present day actually find is a highly uneven or jerky record; that is, species appear in the sequence very suddenly, show little or no change during their existence in the record, then abruptly go out of the record.[2]

They are not negligible gaps. They are periods, in all the major evolutionary transitions, when immense physiological changes had to take place. A couple of examples will do for the moment.[3]

One, how did fish become amphibians? The most important body changes here are that fins must turn into feet, and a pelvis must develop to support the amphibian's weight. (There are many associated changes of enormous complexity and difficulty, such as gills being transformed into lungs, but these, being soft tissue, might not show up as fossils.) On the basis of gradual Darwinian evolution, you would expect a wealth of transitional forms showing the development of the appropriate fins, the loss of others, and the slow strengthening of the pelvic bones.

There are none that show a continuous chain or series. The "link" most often suggested is between rhipidistian crossopterygian fish and an amphibian genus known as *Ichthyostega*. A similarity of skull pattern, and of vertebrae, can be seen; and the fin-bones in crossopterygian fish are also said to be the forerunners of amphibian limbs. But the creatures are aeons apart, anatomically. The limbs and pelvic girdle of the amphibian are well adapted for walking on land. Yet there is nothing whatsoever in the fossil record to show how they reached this stage.

Two, how did mammals evolve their jaw? The change between the reptile jaw and the mammal jaw is profound: and if orthodox evolutionary teaching is correct, the first is ancestral to the second.

All reptiles have at least four bones on each side of the lower jaw, and one bone in each ear. With mammals, on the other hand, the position is almost precisely reversed—every known mammal, alive or extinct, has a single jaw bone and three bones in each ear.

These bones are readily preserved as fossils. Yet there are no existing fossils of transitional forms showing, for instance, three or two jaw bones, or two ear bones. No one has put forward an explanation, either, of how a transitional form might have managed to chew while its jaw was halfway through being re-articulated, nor how it would hear while two jaw bones were being absorbed in its ear structure.

These are severe but by no means untypical conflicts between palaeontology and Darwinism. Professor N. Heribert-Nilsson of Lund University, Sweden, after forty years in the subject, summed up in his book *Synthetische Artbildung*:

It is not even possible to make a caricature of evolution out of palaeobiological facts. The fossil material is now so

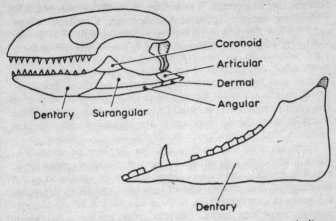

In the reptile–mammal transition, no fossils exist to indicate how the four bones in the reptile jaw (above) were transformed into the single bone of mammals (below)—a puzzle for the world's natural history museums.

complete that the lack of transitional series cannot be explained by the scarcity of the material. The deficiencies are real, they will never be filled.[4]

So Darwinian theory, at least in its original form, runs into apparently crucial difficulties even at the first hurdle. We look to the fossils to show us what happened in the course of evolution, and the key ones are not to be found.

What is the explanation?

First, some non-explanations.

Professor Stephen Jay Gould of Harvard University, a vigorous supporter of Darwin but an opponent of many cherished evolutionary dogmas that have grown up around him, calls fossil gaps "the trade secret of palaeontology." Reading popular or even textbook introductions to evolution, one sees what he means: you might hardly guess that they exist, so glibly and confidently do most authors slide through them. In the absence of fossil evidence, they write what have been termed "just so" stories. A suitable mutation just happened to take place at the crucial moment, and hey presto, a new stage of evolution was reached.

Here, for instance, is David Attenborough, a zoologist whose book *Life on Earth* was based on a popular BBC-TV series, explaining away what happened during the vital 100 million years when fish were evolving. After agreeing that the fossil record permits "only a brief isolated glimpse of the progress of the invertebrates," he is also confident enough to describe, as if they were demonstrable, the following chain of events:

> The corals arrived and began to build reefs, and the segmented animals developed into forms that soon would leave the sea and establish a bridgehead on land. Important changes also took place among the proto-fish. The slits in the sides of their throats, which had originated as filtering mechanisms, were walled with thin blood vessels so that they also served as gills. Now the pillars of flesh between them were stiffened with bony rods and the first pair of these bones, slowly over the millennia, gradually hinged forward. Muscles developed around them so that the front ends of the rods could be moved up and down. The creatures had acquired jaws. The bony scales in the skin which covered them grew larger and sharper and became teeth. No longer were the backboned creatures of the sea lowly sifters of mud and strainers of water. Now they could bite. Flaps of skin grew out of either side of the lower part of the body, helping to guide them through the water. These eventually became fins. Now they could swim. And so, for the first time, vertebrate hunters began to propel themselves with skill and accuracy through the waters of the sea.[5]

Such accounts are really not much more helpful than a line of plausible patter before a conjurer says abracadabra and produces a rabbit (or in the above case, a fish) from his hat. Some writers admit this—for instance N. J. Berril in *The Origin of Vertebrates:*

> There is no direct proof or evidence that any of the suggested events or changes ever took place; what strength the argument may have comes from whatever wealth of circumstantial detail I have been able to muster. In a sense this account is science fiction, but I have myself found it an interesting and enjoyable venture to speculate concerning the Cambrian and pre-Cambrian happenings that may have led to my own existence.[6]

Other writers go even further and take a line of honest bafflement— for instance, F. D. Ommaney in *The Fishes:*

> How this earliest chordate stock evolved, what stages of development it went through to eventually give rise to truly fishlike creatures, we do not know. Between the Cambrian, when it probably originated, and the Ordovician, when the first fossils of animals with really fishlike characters appeared, there is a gap of perhaps 100 million years which we will probably never be able to fill.[7]

Or A. S. Romer in *Vertebrate Paleontology,* writing about rodents. These animals, the most profuse on Earth today (in number of species, they exceed all other mammals combined), flourish under all conditions. But their origin seems beyond explanation:

> When they first appear, in the later Palaeocene, in the genus *Paramys*, we are already dealing with a typical, if rather primitive, true rodent, with the definitive ordinal characters well developed. Presumably, of course, they had arisen

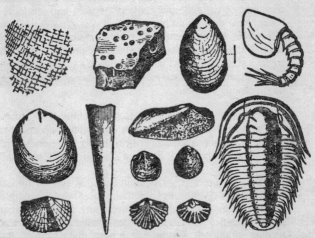

A selection of Cambrian fossils. The trilobite (lower right) is the most complex, but for all these creatures the mystery is the same: why are there no transitional fossils leading up to them?

from some basal, insectivorous, placental stock; but no transitional forms are known.[8]

Or any number of writers puzzling over the explosion of life forms at the beginning of the Cambrian period. Here, in the space of about ten million years, a curtain was raised on a stage teeming with living things. After 3,000 million years in which nothing more complicated than bacteria and slime lived upon our world, came the dawn of life. Billions upon billions of fossils have been found, showing a marine life that suddenly became rich and abundant: clams, snails, octopuses, crustaceans with hard shells and jointed legs, spiny-skinned animals such as starfish, sea urchins and sea lilies. The dominant life form was the now-extinct sea creature known as a trilobite, up to a foot long, with a distinctive head and tail, a body made up of several parts, and a complex respiratory system.

But although there are many places on Earth where 5,000 feet of sedimentary rock stretch unbroken and uniformly beneath the Cambrian, not a single indisputable multi-celled fossil has been found there. It is "the enigma of palaeontological enigmas" according to Stephen Gould. Darwin himself said he could give "no satisfactory answer" to why no fossils had been discovered (panel 1). Today's scientists are none the wiser:

> A century of intensive search for fossils in the pre-Cambrian period has thrown very little light on this problem.[9]

> When we turn to examine the Precambrian rocks for the forerunners of these Early Cambrian fossils, they are nowhere to be found.[10]

> The absence of any record whatsoever of a single member of any of the phyla in the Pre-Cambrian rocks remains as inexplicable on orthodox grounds as it was to Darwin.[11]

Explanations

Moving from non-explanations to explanations, there seem to be two main categories, albeit somewhat contradictory. On the one hand we are told that there are "literally thousands of transitional forms, and more are discovered every year."[12] On the other hand it is said that new ideas about the nature of the

PANEL 2
Evolution of the Horse?

The evolution of the horse has been held up by textbooks since the turn of the century as the way in which fossils demonstrate evolution in action. The first classification, a simple one involving four straightforward steps, was made in 1874, and as recently as 1964 the definitive *Atlas of Evolution* by Sir Gavin de Beer, Director of the British Museum of Natural History, contained a neatly graded ladder of horse evolution starting with pictures of primitive and small ones some seventy million years old and finishing with the large, sophisticated horse of today. A number of geologists have spoken almost lyrically about how the horse supports the general theory of evolution—for instance, this address to the Geologists' Association in 1966:

> The beautiful gradational sequence which these fossils show is now so well described that we need only summarize its major features. These involved the increase in body size, the increase in size and shape of the skull, changes in the teeth, involving the premolarization of the molars, and the deepening of the teeth from low crowned to high

crowned, together with the infilling of the depressions in the upper surfaces with cement. With these were associated changes in the limbs, with the gradual reduction in the number of toes, and in the whole change in posture from pad-footed to spring-footed. Now this series is incontrovertible. It provides clear evidence of the transition of one genus into another over a period of something like seventy million years.[18]

Now maybe the modern horse emerged in this way, or maybe it didn't. But it is emphatically not as straightforward as that in the fossil record—to quote Heribert-Nilsson again, "the family tree of the horse is beautiful and continuous only in the textbooks." Here are some of the problems.

1. A complete series of horse fossils is not found in any one place in the world arranged in rock strata in the proper evolutionary order from bottom to top. The sequence depends on arranging Old World and New World fossils side by side, and there is considerable dispute as to what order they should go in. According to one authority, there are now so

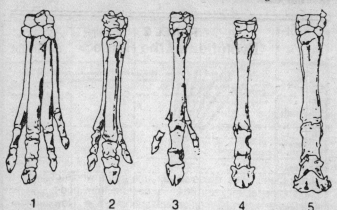

A typical textbook illustration of the evolution of the horse, showing a gradual reduction in the number of toes. But the truth is much more complicated. The various ancestral horses were of all sizes, and the increase and decrease in the number of toes was equally erratic.

many fossil horses competing for a place on the evolutionary tree that "the story depends to a large extent upon who is telling it and when the story is being told."

2. *Eohippus*, supposedly the first horse, doesn't look in the least like one—and indeed, when first found was not classified as such. It is remarkably like the present-day *Hyrax* (or daman), both in its skeletal structure and the way of life that it is supposed to have lived. Moreover, *Eohippus* fossils have been found in surface strata, alongside two modern horses, *Equus nevadensis* and *Equus occidentalis*.

3. Museum displays and textbook evolutionary ladders illustrate only a partial and favorable selection of reconstructed horses. While the general trend seems to have been toward larger horses, the first three supposed horse fossils (*Eohippus*, *Orohippus* and *Epihippus*) actually decline in size. In any case, the range in size of horses alive today, from the tiny American miniatures to the great shire horses of northern Britain, is the same as is found in the fossil record. If some palaeontologist, millions of years hence, arranged them in order from smallest to largest, it might well appear a con-

GEOLOGICAL EPOCH		AFRICA	EUROPE AND ASIA	NORTH AMERICA	SOUTH AMERICA
RECENT	Upper	TARPAN KIANG HORSE ASS ZEBRA QUAGGA	HORSE ASS	HORSE ASS	HORSE ASS
	Middle				
	Lower				
PLEISTOCENE	Upper				
	Middle		EQUUS	EQUUS	EQUUS
	Lower		EQUUS	EQUUS, EQUUS SCOTTI	EQUUS, ONOHIPPIDION
PLIOCENE	Upper		EQUUS	EQUUS	EQUUS, HIPPIDION
	Middle	HIPPARION		PLIO-HIPPUS	HIPPIDION
	Lower	HIPPARION, HIPPARION GRACILIS		PLIO-HIPPARION	HIPPIDION
MIOCENE	Upper	HIPPARION WHITNEYI		PROTO-HIPPUS	HYPOHIPPUS MATTHEWI
	Middle	ANCHITHE-RIUM		MERYC-HIPPUS	HYPOHIPPUS
	Lower			MIO-HIPPUS	HYPOHIPPUS EQUINUS, HYPOHIPPUS, PARAHIPPUS
OLIGOCENE	Upper			MIO-HIPPUS	
	Middle			MESO-HIPPUS	MESOHIPPUS INTERMEDIUS
	Lower			MESO-HIPPUS	MESOHIPPUS BAIRDI
EOCENE	Upper			EPI-HIPPUS	
	Middle			ORO-HIPPUS	
	Lower		HYRACOTHERIUM	EO-HIPPUS	

This chart shows just one of the many conflicting versions of horse ancestry.

vincing example of evolution at work.

4. The sequence from many-toed to one-toed animals is equally erratic, with numerous contradictions and regressions to the theoretically ideal order.

5. Even when all possible fossils are included, there appear to be major jumps in size of the horses from one genus to the next, without transitional examples.

---

evolutionary process mean that very few intermediates were ever turned into fossils, and we are lucky to find the ones we have—"we must accept that the chances of finding the transition from one species to another preserved for us in the rocks, in a statistically significant way, are extremely remote."[13]

It takes a while to realize that the "thousands" of intermediates being referred to have no obvious relevance to the origin of lions and jellyfish and things. Most of them are simply varieties of a particular kind of creature, artificially arranged in a certain order to demonstrate Darwinism at work, and then rearranged every time a new discovery casts doubt upon the arrangement.

The family tree of the horse (panel 2) is a good example of the trouble museums get into when they try this sort of thing. Once portrayed as simple and direct, it is now so complicated that accepting one version rather than another is more a matter of faith than rational choice. *Eohippus*, supposedly the earliest horse, and said by experts to be long extinct and known to us only through fossils, may in fact be alive and well and not a horse at all—a shy, fox-sized animal called a daman that darts about in the African bush.

The "thousands" of intermediates also include a number of creatures of about the same explanatory value as the crossopterygian fish—that is, almost none. They are simply speculative candidates in the evolutionary ladder—disconnected links in a hypothetical chain. All the most popular examples have become discredited, one way or another. The so-called "walking catfish" of Florida, frequently cited as living halfway fish-amphibian creatures, do not, in fact, walk. They slither along on their bellies using the same motion as when they swim. (Actually, they climb trees too. It might as well be argued that they are therefore intermediate to birds.)

For years, also, the primitive bony fish called the coelacanth, which was abundant 400 million years ago, was quoted as being an intermediate, because of the way its fins had certain limblike characteristics. It was supposed to have been capable of lurching forward on land in search of food, eventually staying there for longer and longer periods until, seventy million years ago, it disappeared from the fossil record.

However, now that several dozen coelacanths have been caught off the coast of Madagascar, all apparently unchanged from their ancestral form, perfectly adapted to their natural deep sea habitat, and showing no signs of lurching about on land, they have been quietly dropped from the textbooks as transitional forms.

A similar fate has happened to the lungfish, which until quite recently was regularly held up as a transitional creature, being able to breathe oxygen into its bloodstream through a primitive pair of lungs. But it has lived like that, obstinately refusing to evolve, for 350 million years, and biologists, having reexamined the bone structure in its head, now discount the possibility that its descendants could ever have turned into amphibians.

Even the most famous of all intermediates, *Archaeopteryx*, the "diminutive winged dinosaur" is in difficulties (panel 3). An examination of the feather structure seen in the fossils, undertaken at the Smithsonian Institution in Washington, established that it was the same as that used by modern birds. "*Archaeopteryx* did fly," reported the *New Scientist*.[14] The creature may very well not have been a dinosaur at all, but a strange kind of ancient bird.

A Statistical Solution

So the "thousands" of intermediates, including the best-known and most widely quoted, do not lead us very far, nor explain away the fossil gaps. We are left then with the statistical argument, which is altogether more sophisticated and subtle. It takes us a long way from what Darwin thought, and from what we have been traditionally taught, but for all that it is the most persuasive explanation that has so far been put forward. It goes something like this.

It is a mistake to look for intermediate forms among the vertebrates and then complain because you can't find them, because this is precisely where the fossil record is least adequate. Any number of events may happen to prevent the fossil of a highly developed creature being formed, or found. It is a matter

PANEL 3
Is *Archaeopteryx* an Intermediate?

Fortuitously for Darwin and his theory, it seemed, *Archaeopteryx* was discovered in limestone rocks in Bavaria in 1860 less than two years after *The Origin of Species* was published. First, the outline of a feather was seen, giving rise to its name, which means simply "ancient bird." A year later, in a nearby quarry, an almost complete skeleton was found, its wings outstretched, with a clear imprint of feathers on them.

Its importance, then and now, was that *Archaeopteryx* appeared in the same strata as dinosaur fossils, and appeared at first sight almost as much a reptile as a bird—"a providentially timed confirmation of Darwin's proposition that one group of animals developed into another by way of intermediate forms," it was recently suggested. Biologists as a whole regard it as authoritative evidence of Darwinian evolution at work. "It proved beyond any argument" that there existed an animal with both reptilian and bird features, according to one account "even today, there is no more convincing example" of a transitionary link, says another.

But is the case for *Archaeop-*

teryx quite so unambiguous as these claims make out? Apparently not. Every one of its supposed reptilian features can be found in various species of undoubted birds.

1. *It had a long bony tail, like a reptile's, on which feathers grew.*

While it is generally true that reptiles have tails, and birds appear not to, the detailed position is more complex. In embryo, some living birds have more tail vertebrae than *Archeaopteryx* does, which later fuse to become an upstanding bone called the pygostyle. The bone and feather arrangement on a present-day swan shows striking similarities to *Archaeopteryx*. According to one authority, there is no difference in principle between the ancient and modern forms: "the difference lies only in the fact that the caudal vertebrae are greatly prolonged. But this does not make a reptile."

2. *It had claws on its feet and on its feathered forelimbs.*

But so do some modern birds, such as the hoatzin in South America and the touraco in Africa. The ostrich of today, which also has three claws on its wings, has been suggested by some experts to have more supposed reptilian features than *Archaeopteryx*—but nobody, of course, considers the ostrich a transitional form.

3. *It had bony jaws lined with teeth.*

Modern birds do not have teeth. But many ancient birds did, particularly those in the Mesozoic, and there is no suggestion that these are intermediates. It is just as convincing to argue that *Archaeopteryx* was an early bird with teeth.

4. *It had a shallow breastbone that would have given it a feeble wing beat and poor flight.*

Modern woodcreepers such as the hoatzin have similarly shallow breastbones, and this does not disqualify them from being classified as birds. And there are, of course, many species of bird, now and in the past, which are incapable of flight.

In any case, recent examination of *Archaeopteryx*'s feathers at the Smithsonian Institution has shown that they are the same as those belonging to many modern accomplished fliers. "This implies at the very least that the beast could glide at some speed and lays to rest the notion that the feathers evolved as either heat insulation or as an aid to trapping insects."

5. *Its bones were solid, like a reptiles's, not thin or hollow, like a bird's.*

Another idea that has been drastically revised. The long bones of *Archaeopteryx* (wings, legs) are known now to have been both thin *and* hollow. It is still debated whether they were "pneumatized" like a bird's, i.e., containing an air sac.

6. *It predates the general arrival of birds by sixty million years.*

Until 1977, *Archaeopteryx* was uniquely early in the fossil record. But in that year, archaeologists from Brigham Young University discovered, in western Colorado, a fossil of an unequivocal bird, in rocks of the same period as *Archaeopteryx*. Professor John H. Ostrom of Yale University, who positively identified the specimen, commented: "It is obvious we must now look for the ancestors of flying birds in a period of time much earlier than that in which *Archaeopteryx* lived."

This discovery much weakens the case for *Archaeopteryx* as an intermediate, and makes it that much more likely that the creature was just one of a number of strange birds living at that time. Professor Heribert-Nilsson commented forcefully that "they are no more reptiles than the present day penguins with their wing-fins are transitional forms to fish."

The further point might be made that even if *Archaeopteryx* is in fact a halfway form from reptiles to birds, it is still not very enlightening about the process of evolution, nor in any way evidence of Darwin's hoped-for gradual transitions. For that, we would have to see in the fossil record the slow development of feathers (perhaps from scales, perhaps from some other origin) and the hierarchical change of amphibian dinosaurs into delicate, light-boned creatures that could soar above the Earth. And here, characteristically, the rocks are mute.

of chance if a plant or creature dies in a place with the right amount of lime to preserve it. Rock formations have folded over on top of themselves and eroded, so that fossils show up irregularly. Many fossils lie at the bottom of the world's oceans, out of our grasp.

Above all, soft parts of a creature tend not to be preserved. Thus in the reptile–mammal transformation, in the Triassic period, there are many "mammallike reptiles" which might be one or the other. But it is difficult to say certainly, because the essential features of a mammal (hair, warm blood, suckling the young, giving live birth, etc.) are not such as to be fossilizable.

Moreover, these fossils are extremely rare. Many palaeontologists go through their entire career without finding one, let alone showcase examples such as a coelacanth or a horse. Instead, it is much more profitable to seek the pattern and means of evolution in the vastly more abundant, if less ostentatious, fossils of marine invertebrates, particularly the various kinds of shellfish whose history can be traced through tens of millions of years.

Here evolution can truly be "seen," as shells gradually change shape, and one variety of creature supplants another. In a sense, there is a superabundance of intermediates, for no species can be seen to last in exactly the same form forever. A Jurassic oyster of 150 million years ago may have tasted the same as one of the varieties in the world today, but it is no longer alive; it is a distinct and readily recognizable fossil which then passed through many intermediate forms prior to our present-day species.

Indeed, looking at the picture of life on Earth as a whole, you can see that of all the hundreds of thousands of species of plants and animals that exist now, hardly a single one has been found absolutely unchanged as a fossil in rocks older than about ten or twenty million years, which is no time at all geologically. Transitional types are *not* invariably lacking, but the further back in the record you go, the less evident they are. George Gaylord Simpson, when he was Professor of Vertebrate Palaeontology at Harvard, summarized the position like this:[15]

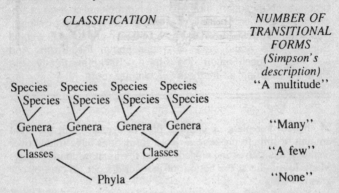

CLASSIFICATION	NUMBER OF TRANSITIONAL FORMS (Simpson's description)
Species Species Species Species	"A multitude"
Genera Genera Genera Genera	"Many"
Classes Classes	"A few"
Phyla	"None"

Professor Ernst Mayr, also of Harvard, and like G. G. Simpson one of the dozen or so most redoubtable evolutionary theorists this century, proposed a pattern for evolution which made it even more unlikely that intermediate fossils would be found. He called it *allopatric speciation,* which means literally that new species are created "in another place." He supposed that a new mountain range, or some other geological/meteorological event, would cause a small portion of a population (animal or plant) to become isolated. For various genetic reasons, which we shall come to in the next chapter, evolution in these circumstances would take place more rapidly, and natural selection operate more intensively, and a new, more powerful species would emerge. Having done so, it would spread back into its former territory and quickly conquer it. Naturally, the chance of finding fossils of this peripheral group while it was undergoing the process of speciation would be extremely small. Generally, you would simply find the finished form when it had proved itself successful—and hence the reason for the fossil gaps.

The familiar "tree of life" has changed substantially over the years as scientists have come to accept that there are gaps in the fossil record that may never be filled. (a) Professor Ernst Haeckel, a contemporary of Darwin, produced this version at the turn of the century.

PEDIGREE OF MAN.

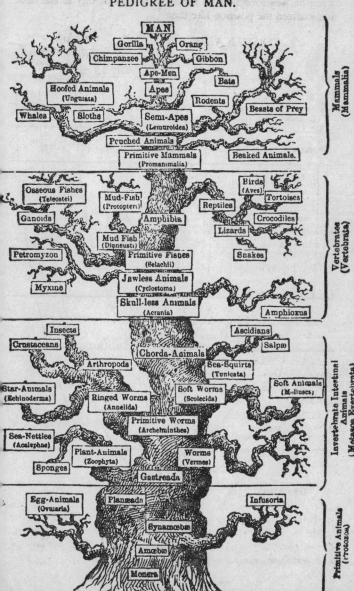

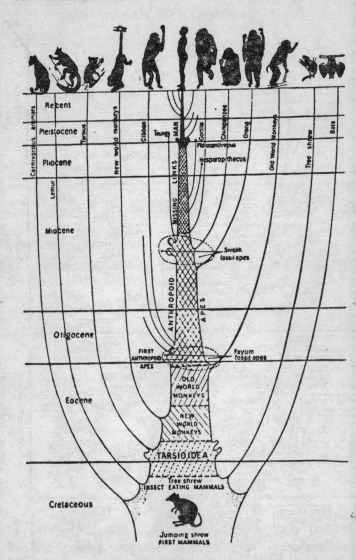

(b) A textbook example of the 1920s

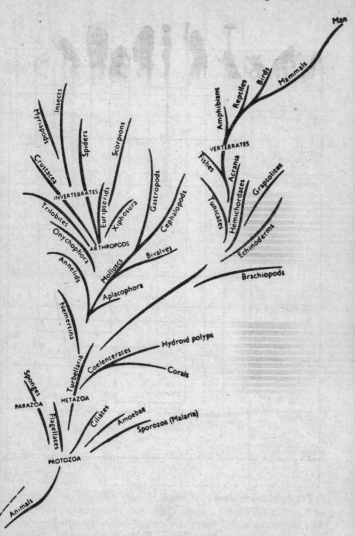

(c) Sir Gavin de Beer's Atlas of Evolution *represented academic thinking of the 1950s; the impression of a tree is given, but the "branches" are left unjoined.*

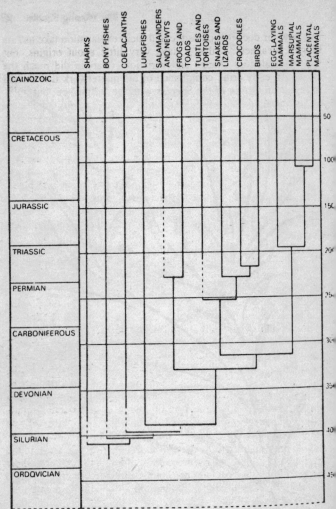

(d) The British Museum of Natural History's Evolution handbook published in 1978 gave a stylized and even more cautious version. The caption warned: "Solid vertical lines show groups known by fossils; dotted lines indicate 'gaps' in the fossil record—periods when the group is inferred to have existed, but no fossils have been found. These gaps might be due to failures in fossilization, or to mistakes in the genealogy, or to wrongly identified fossils; or they could be (and have been) taken to show that the theory of evolution is wrong."

But how far does this plausible and lucid explanation take us? Its biggest weakness is that it says very little about origins, but simply catalogues change and variation. (Darwin did much the same. An early critic complained of his masterwork that he had written "an *Origin of the Species* with the 'origin' cut out.")[16]

Technically, a species is a type of plant or creature that cannot reproduce with a different type, and in this sense the various types of oyster that have supplanted each other over the ages are certainly different species, which have possibly evolved through a process of natural selection (it is less than obvious that the new types are an improvement on the previous types). But we are still none the wiser about how the oyster got there in the first place. Moreover, belief in allopatric speciation embraces the confession that we never *will* find the hardware of evolution—we shall always have to rely on indirect and less satisfactory proofs, such as statistical ones.

Indeed, if it is really true that there are "thousands" of intermediates, and "more are discovered each year," it is curious that evolutionists seem to be growing less sure, rather than more sure, about how one thing evolved into another. The "trees of life" that adorn textbooks and museum displays are nothing like as confident as they were. Up to about 1950, the trunks, branches and twigs were almost always joined together, demonstrating the supposedly smooth progression of life forms from a common origin. Then, gaps and dotted lines began to appear where the "missing" fossils ought to be, although a strong impression was still given that the tree of life was still a tree. Lately, even that supposition seems in the process of being gently dropped (see the section on cladistics in Chapter Eight). Plants and creatures are simply shown as having "arrived" in the evolutionary record, and the question of their ancestry ignored.

Puzzling though it is to those of us brought up to believe in the gradual ascent of man from amoeba, the fossil record contradicts this. "Evolution requires intermediate forms between species and palaeontology does not provide them," wrote a contributor to *Evolution* in 1974.[17]

It is a major Darwinian impasse.

On its own, the strangely absent evidence may not spell the death of Darwinism; we could perhaps just call it one strike against. But moving away from the museums and into the biology laboratories, does the theory fare any better?

PANEL 4
Do Fossils Show Gradual Change?

One of the most contentious evolutionary disputes in recent times has surrounded the argument that the fossil record shows stasis, rather than change—that once a new species has become established on Earth, it remains there substantially unaltered in shape until its abrupt extinction some millions of years later. It is an idea which surfaces repeatedly in this book, and it has an immediate bearing on gaps in the fossil record, and Darwinian attempts to explain them.

For it is an undoubted fact, as Darwin noted many times, that the fossil record is extremely imperfect. It is unreasonable to expect, as evidence of gradual Darwinian evolution, an unbroken sequence of fossils showing the emergence of a modern creature from a primitive ancestral form—fossils are too rare and too patchily distributed for that.

However many—perhaps most—palaeontologists today have become convinced that the pattern is altogether different. There are no intermediate fossils showing a three-quarter-length giraffe neck, for instance. While this may be due to an inadequate sample (panel 26), the same cannot be said for the multitude of fossils of bivalve molluscs (clams, oysters, etc) which have been observed, counted and analysed statistically.

Professor Anthony Hallam, of the Department of Biological Sciences at Birmingham University, England, told a conference on evolution in Chicago in 1980 that his studies of Jurassic bivalves showed emphatically that over periods of 10–12 million years they changed hardly at all, except in size. After this period of stasis they became, or were replaced by, a markedly different species.

The growing realization that a pattern of stasis and extinction is the rule rather than the exception shows not just that the fossil gaps are real (which more or less everybody now accepts); but also that they are highly mysterious, significant, and very unhelpful to Darwinism.

CHAPTER TWO

Natural Limits

The laws governing inheritance are quite unknown; no one can say why the same peculiarity in different individuals of the same species, and in individuals of different species, is sometimes inherited and sometimes not so; why the child often reverts in certain characters to its grandfather or grandmother or other much more remote ancestor; why a peculiarity is often transmitted from one sex to both sexes, or to one sex alone, more commonly but not exclusively to the like sex.

—Charles Darwin,
The Origin of Species

Not long after *The Origin of Species* was published, and just at the time when Darwinism was becoming respectable among younger scientists, the theory seemed to have been dealt a lethal blow. In 1867 Fleeming Jenkin, a Scottish professor of engineering, homed in on Darwin's greatest weakness: his ignorance, which he admitted, as to why some characteristics are passed on to the next generation, and some aren't.

His theory demanded that a favorable new trait—longer legs, or better eyesight, for instance—should be inherited by the creature's offspring, and then by their offspring in turn, and so on. The current belief, which Darwin implicitly accepted in his book, was that the characters of each parent were "blended" in the children.

Fleeming Jenkin pointed out that this couldn't happen—or at least, it was so improbable as to make no sense. Unless both parents had longer legs, the offspring would have only half a chance of inheriting the newly lengthened legs, and their offspring only half again, and so on until within a very few generations the original variation would be blended away to insignificance.

As Darwin grew older, he became increasingly unsure that natural selection alone was the answer to evolution. In 1882 he died without knowing that this particular problem—of "fixing" favorable characteristics—had already, seemingly, been overcome. The apparent solution lay in the archives of an obscure natural history society in Brno, Czechoslovakia. Ironically, Darwin could have learned of it in his lifetime, and saved himself much mental agonizing, for it was in 1866 that Gregor Mendel, a monk who taught science and mathematics in the town's high school, presented his definitive paper.

He described his experiments in breeding varieties of sweet peas. These demonstrated, for the first time, that there were definite hereditary units which passed down specific characteristics—in this case, the color of the peas—and that these remained undiluted from generation to generation. He called them *factors;* nowadays they are known as *genes*.

Just how genes do their work is far more complex than Mendel could have imagined. It was not, as he thought, a simple matter of having one gene for blue eyes, another for blond hair, and so on. Instead, genes may work singly or in blocks; may interact with one another; may lie dormant until awakened by some stimulus in the environment, or by sexual mixing; and vast amounts of genetic material seem to have no useful or known function (in humans about ninety percent is surplus to requirements and is technically described as "redundant" because scientists don't know what it is for).

The relentless flow of papers on molecular biology in specialist journals—in any week, there are several dozen giving results of new experiments—deal with the dauntingly complicated details of the processes that are thought to go on at this submicroscopic level. The basic principle, however, is both simple and, on first acquaintance, intellectually satisfying (which is why, like Darwin's notion of natural selection, it has so easily captured public acceptance).

Life's Building Blocks

The science of genetics, it is said, describes the mechanics of life itself. It explains why none of us (except identical twins) are exactly alike, but are at the same time fairly similar—human beings are one species, monkeys another, and jellyfish something completely different.

It is because virtually all life is made up of building blocks known as cells (humans have about 100 million million cells), and within the nucleus of each cell are genes—a code of information that is unique to each plant and creature.

Cells renew themselves constantly. The genetic code instructs them how to do so identically, time after time. We die with the same set of fingerprints we were born with; genes have maintained the individual pattern of skin cells through our life.

The genes that determine the nature and task of each cell are long strands of the chemical DNA, coiled like a spiral staircase in the famous double helix. A chain's links are made up of just four kinds of acid molecules, described scientifically by their initial letter (A for adenine, T for thymine, C for cystosine and G for guanine). Thus DNA for one species might look like this:

and so on for a million or more links, and DNA from another species, with the same four acid molecules (called nucleotides) but in a different order, like this:

and so on for a million or more links.

Had Darwin lived to know it, the importance of genes to this theory is that genes in the germ cells can be inherited. During reproduction, the genetic message of one living thing is mixed with the genetic message of another living thing of the same species, and half of each is inherited by the offspring. And even though Mendel was oversimplistic in imagining a one-for-one relationship between a gene and an outward characteristic, basically he was right. Gene patterns and gene complexes are passed on unchanged from mother to child, which is why a child has the "looks" of its parents.

So, shortly after the turn of the century, Darwin's theory suddenly seemed plausible again. Next, it was found that once in a while, absolutely at random (about once in ten million times during cell division, we now know) the genes make a copying mistake. These mistakes are known as mutations, and are mostly harmful. They lead to a weakened plant, or a sick or deformed creature. They do not persist within the species, because they are eliminated by natural selection. Other mutations, it is thought, are neutral in effect, and lie dormant within the genetic system.

However, followers of Darwin have come to believe that it is the occasional beneficial mutation, rarely though it happens, which is what counts in evolution. They say these favorable mutations, together with sexual mixing (panel 5), are sufficient to explain how the whole bewildering variety of life on Earth today originated from a common genetic source. The theory is that a chance favorable mutation gradually spreads through a population of plants or animals by a process of natural selection of the fittest; and over geological periods of time, a new species emerges. Genetics provides the mechanism that supports Darwin's original insight.

Modern Darwinism

It is fair to say that this explanation of evolution (called the synthetic theory, or neo-Darwinism, because it combines traditional Darwinism, Mendelian inheritance, and the mathematics of population change) has utterly dominated biological science for the last fifty years. The teaching of evolution in virtually all colleges in the western world means the teaching of population genetics.

On either side of the Atlantic, the most senior professors leave the impression that there is no need to look further than this for the origin of life forms. Ernst Mayr of Harvard University: "The proponents of the synthetic theory maintain that all evolution is due to the accumulation of small genetic changes, guided by natural selection."[1] John Maynard Smith of Sussex University has written emphatically that neo-Darwinism's theoretical basis is "necessary and sufficient to account for the evolution of life on this planet to date."[2]

With such apparent unanimity in the textbooks and the classrooms, it comes as something of a surprise to discover that, to almost everybody *except* the population geneticists, the synthetic theory is as full of holes as the fossil record.

Thus Stephen Gould (also of Harvard, but a generation younger than Ernst Mayr, and who approaches the subject via geology and the history of science in addition to biology) confessed in 1980:

I well remember how the synthetic theory beguiled me with its unifying power when I was a graduate student in the

PANEL 5
Changing the Genetic Message

Mutations are caused by natural radiations (such as x-rays and ultraviolet light) and by manmade chemicals hitting and damaging the nucleotide links of DNA. Nucleotides may thus be transformed to other nucleotides (right):

or they may be chemically altered to a form that is not one of the four standard nucleotides:

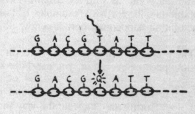

or they may even be taken right out of the chain:

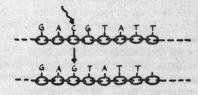

In sexual reproduction, the mother's set of genes is mixed with the father's set of genes in a process that is known as *recombination*. Whole lengths of genes, rather than single letters, are involved:

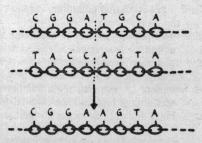

mid-1960s. Since then I have been watching it slowly un-ravel as a universal description of evolution. The molecular assault came first, followed quickly by renewed attention to unorthodox theories of speciation and by challenges at the level of macroevolution itself. I have been reluctant to admit it—since beguiling is often forever—but if Mayr's characterization of the synthetic theory is accurate, then that theory, as a general proposition, is effectively dead, despite its persistence as text-book orthodoxy.[3]

So who is right? Is neo-Darwinism really dead, or dying? The starting point, as with the fossil evidence, must be the showcase exhibits—the textbook examples that are held up to demonstrate evolution happening, here and now.

There are just three places where, in spite of the interminably slow pace of evolution, you can hope to glimpse it in action:

1. In the natural world around us, rapidly changing life forms may indicate the way things evolve.
2. Artificial breeding of animals and plants may show us the evolution of new types.
3. Long periods of evolution can be simulated by irradiating laboratory animals and making them mutate faster.

All three approaches have been tried repeatedly.
Of the first, the peppered moth *Biston betularia* is easily the most widely cited case. It has been studied in its natural habitat in northern Britain since 1849, and over the years biologists have noted with pleasure that the proportion of dark-colored to light-colored moths in the population has varied directly with the amount of industrial pollution near big cities such as Manchester.

The peppered moth is a striking example of evolution in action. The original silver form is well camouflaged on lichen-covered tree trunks but where air pollution has killed the lichen and blackened the tree trunks with soot, a black variety of peppered moth has become increasingly popular and almost replaced the original form in some places. The black mutant had begun to increase in the latter half of the nineteenth century, but, sadly for Darwin, no one knew it at the time. This was just the evidence he needed to show the effectiveness of natural selection. Many other insect species have now been found to have black varieties in areas with air pollution; the phenomenon is known as industrial melanism.[4]

The rise in the population of dark varieties of the peppered moth was indeed dramatic: from about one percent of the population in 1849 to ninety-eight percent in 1900. Once the camouflage disappeared, birds could easily see and eat the silvery variety. Industrial melanism has now been established in more than 100 kinds of insects in Britain, and there are many other examples in Europe and North America.

Evolution is also said to have been demonstrated in front of our eyes by mutant strains of creatures that have become resistant to antibiotics or poisons. The bacteria *Escherichia coli*, living in our stomachs, has produced a number of variants that are immune to earlier antibiotics such as streptomycin. Various strains of housefly have survived the onslaught of DDT. Rats have defeated all attempts to exterminate them with poison. Some rabbits failed to succumb to myxomatosis.

Micro-evolution and Macro-evolution

So far as living showcase examples of evolution are concerned, that is about it. As we noted earlier, evolution takes place on such a huge timescale that we are destined never to see anything major happening within our own lifetime. The best we can do is to look at examples of change within living species, and then make the assumption that if the process went on long enough, evolution would happen in a Darwinian way.

It is a big assumption. Nobody disputes that natural selection has a role to play in the kind of *micro-evolution* we can see happening here—a small amount of adaptive change within a particular kind of living thing. You may even arrive at what is technically a new species, in the sense that the newcomer cannot reproduce with its forebears. Sometimes you can identify the gene substitution that accompanies the change.

But is *macro-evolution* the same mechanism writ large? A strictly neutral interpretation of the cases above might conclude simply that in favorable conditions some living things multiply, and in unfavorable conditions they die. The "new" bacteria, houseflies, rabbits, and so on, are nothing more than existing varieties, which simply multiplied to fill the ecological gap left when the dominant forms were killed off by something new and harmful.

The peppered moths, after all, stayed from beginning to end *Biston betularia*. Now that smoke-control policies are operating in the Manchester area, and the air is becoming cleaner, the original silvered variety of moth is on the increase again. Nothing, therefore, has changed except the ratio.

Artificial Evolution

So what of the second area of evidence, selective breeding? Darwin made much of this. It was one of the insights of *The Origin of Species* to recognize that nature in the wild might encourage "improvements" in the same sort of way as the farmer breeds better and larger pigs, horses, cattle and the like. He devoted the whole of the first chapter of his book to the subject, and returned to it many times.

Since then there have been very large stockbreeding successes. Grain, beans, peas, fruit, milk, wool, beef, eggs—it is hard to think of a domesticated plant or animal that has not been improved in one respect or another during the last century.

However, so far as evolution is concerned, it is now absolutely clear that there are firm natural limits to what can be done. Remarkable achievements can be made by crossbreeding and selection inside the species barrier, or within a larger circle of closely related species, such as the wheats. But wheat is still wheat, and not, for instance, grapefruit. Between 1800 and 1878, the sugar content of beets was raised from six to seventeen percent. A half century of further breeding failed to make any difference.

Although Darwin and his successors have invoked these and other examples as evidence of evolution, breeders with practical experience flatly disagree. Luther Burbank, perhaps the most famous plant breeder in the history of the United States, once pointed out that nobody had succeeded in growing black tulips or blue roses, because the genetic material was simply not there. "I know from experience that I can develop a plum half an inch long or one two-and-a-half inches long, with every possible length in between, but I am willing to admit that it is hopeless to try to get a plum the size of a small pea, or one as big as a grapefruit. I have roses that bloom pretty steadily for six months of the year, but I have none that will bloom twelve, and I will not have. In short, there are limits to the development possible."[5]

The reason for reaching these evolutionary dead ends is not hard to find: the genetic code in each living thing has its own built-in limitations. It seems designed to *stop* a plant or creature stepping too far away from the average. To use a somewhat oversimplified, but graphic, illustration, imagine a box filled with marbles of different sizes (representing the genes). It might take you some time to sort out the biggest ones. But once you have done so, your search is complete. The marbles will not, of their own accord, grow bigger; and nor will a creature that has been bred to inherit all the "big" genes.

Although this simple arithmetical approach has been rather overtaken by recent advances in genetics, which show that genes tend to interact among each other in a startling and unpredictable way, the principle holds good in practice. Every series of breeding experiments that has ever taken place has established a finite limit to breeding possibilities. Genes are a strong influence for conservatism, and allow only modest change. Left to their own devices, artificially bred species usually die out (because they are sterile or less robust) or quickly revert to the norm.

Nor have these experiments simply happened during the last century or so. Ever since Neolithic man started crossbreeding wild varieties of grass, such as einkorn, more than 10,000 years ago, plant domestication can be seen in the archaeological record. Domesticated dogs arrived about the same time. Yet in the whole of this time period, there is no hint of wheat or dogs changing into anything except different kinds of wheat and dogs.

Monstrous Mutations

So does the third category of evidence offer any better indication of evolution at work? Can speeded-up mutations demonstrate how one kind of creature turns into another?

The geneticist's favorite laboratory animal is the common fruit fly *Drosophila melanogaster*. Since the early 1900s, starting with the work of the biologist T. H. Morgan, tens of millions of flies have been involved in experiments. They have been selectively crossed with each other, and bombarded with various strengths and frequencies of x-rays so that the mutation rate has

PANEL 6
Can a Fruit Fly Be Artificially Improved?

Since early this century, laboratory researchers have been trying to create, through irradiation and selective breeding, a fruit fly which has a selective advantage over the wild type, as defined by leaving more offspring of its own kind under certain conditions. Only two have been suggested with any confidence. They are known, after their discoverers, as Rendel's yellow fly and Kalmus's ebony fly.

The first, in certain laboratory conditions, outlives the parent strain, and is cited as an example of a new species emerging because females of the wild type generally refuse to mate with it. The second, again in the laboratory, is more resistant to drought, insecticides, and low temperatures.

However, it is far from clear that in the wild, rather than in controlled conditions, these small differences would be perpetuated through natural selection. The yellow flies are less resistant to drought and insecticide, and might be more vulnerable, because of their bright color, to predators such as birds, frogs, and other insects. The black flies, in turn, are less resistant to high temperatures, are less fertile (except in drought, etc.) and again are more conspicuous than the wild variety. As evidence of helpful mutations, they are, to say the least, ambiguous.

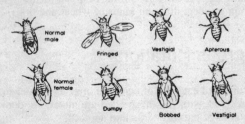

A few of the abnormalities achieved by laboratory breeding of the fruit fly Drosophila. *Despite a mutation rate much increased by exposure to radiation, the flies have not evolved into a different kind of insect.*

sometimes been 150 times faster than normal. Passing under the microscope have been flies with elongated or shortened bodies and limbs, monstrously wrinkled wings, a leg instead of an eye, and so on.

In terms of our knowledge about the workings of DNA and the chromosomes within which DNA resides, the gains from this research have been immense. But applied to theories of evolution the results are as frustrating as in any other kind of selective breeding.

For in spite of the enormously increased mutation rates, all the fruit flies have remained fruit flies. Indeed, out of the millions of mutation, *only two* are arguably "fitter" than the parent stock of flies, and even this is strongly debated (panel 6).

Ernst Mayr, who remains convinced that small-scale gene substitution is the answer to evolution, conducted one striking piece of research on *Drosophila* which, ironically, seemed to demonstrate the opposite.

He selectively bred successive generations of flies to try to increase or decrease the number of bristles they grew, normally averaging thirty-six. He reached a lower limit, after thirty generations, of twenty-five bristles; and an upper limit after twenty generations, of fifty-six bristles. After that the flies rapidly began to die out.

Then, Mayr brought back nonselective breeding, letting nature take its course. Within five years, the bristle count was almost back to average.

This resistance to change has been given the label *genetic homeostasis* and elsewhere in the literature there is an even more mysterious example.[6] In a remarkable series of experiments, mutant genes were paired to create an eyeless fly. When these flies in turn were interbred, the predictable result was offspring that were also eyeless. And so it continued for a few generations.

But then, contrary to all expectations, a few flies began to hatch out with eyes. Somehow, the genetic code had a built-in repair mechanism that reestablished the missing genes. The natural order reasserted itself.

On the face of it, then, the prime function of the genetic system would seem to be to resist change: to perpetuate the species in a minimally adapted form in response to altered conditions, and if at all possible to get things back to normal. The role of natural selection is usually a negative one: to destroy the few mutant individuals that threaten the stability of the species.

Yet ardent neo-Darwinists see it quite differently. Here is a clarionlike affirmation of faith from Sir Julian Huxley, grandson of Darwin's friend Thomas, who until his death in 1975 was always a most outspoken and defiant proponent of the evolutionary synthesis:

> "That is all very well," you may say. "It seems to be true that natural selection can turn moths black in industrial areas, can keep protective coloration up to a mark, can produce resistant strains of bacteria and insect pests. But what about elaborate improvements? Can it turn a reptile's leg into a bird's wing, or turn a monkey into a man? How can a blind and automatic sifting process like mutation produce organs like the eye or the brain, with the almost incredible complexity and delicacy of adjustment? How can chance produce elaborate design? In a word, are you not asking us to believe too much?" *The answer is no . . .* [7]

In rather more sober but still expansively self-confident style, one of America's three leading textbooks of introductory biology describes macro-evolution similarly. (It is the whole and only passage on the subject, verbatim. No doubts or contrary arguments about the mechanism of macro-evolution appear anywhere. Stephen Gould commented: "I can't think of any recent event that depressed me more than reading this.")

> Each of the examples of micro-evolution examined, involving shifts in the frequencies of small numbers of genes, could be multiplied a hundredfold from reports in the scientific literature. Biologists have been privileged to witness the beginnings of evolutionary change in many kinds of plants and animals and under a variety of situations, and they have used this opportunity to test the assumptions of population genetics that form the foundations of modern evolutionary theory. The question that should be asked before we proceed to new ideas is whether more extensive evolutionary change, macro-evolution, can be explained as an outcome of these micro-evolutionary shifts. Did birds really arise from reptiles by an accumulation of gene substitutions of the kind illustrated by the raspberry eye-color gene?
> The answer is that it is entirely plausible, and no one has come up with a better explanation consistent with the known

biological facts. One must keep in mind the enormous difference in timescale between the observed cases of micro-evolution. Under natural conditions the nearly complete substitution of the melanic gene of the peppered moth took fifty years. Evolution of the magnitude of the origin of birds usually, perhaps invariably, takes many millions of years. As palaeontologists explore the fossil record with increasing care, transitions are being documented between increasing numbers of species, genera, and higher taxonomic groups. The reading from these fossil archives suggests that macro-evolution is indeed gradual, placed at a rate that leads to the conclusion that it is based upon hundreds of thousands of gene substitutions no different in kind from the ones examined in our case histories.[8]

The bland complacency of this passage takes some swallowing. What sticks in the throat immediately is the dubious assurance that "increasing numbers" of transitional fossils are being documented, that the origin of major new groups almost always takes "millions of years," and that micro-evolution is unquestionably "gradual." Palaeontological evidence seems overwhelmingly to refute this.

But the key statement is that "no one has come up with a better explanation consistent with the known biological facts." In similar vein, Theodosius Dobzhansky of Columbia University, as eminent a geneticist as Mayr, wrote in 1957: "The process of mutation is the only known source of the new materials of genetic variability, and hence of evolution"[9]; and five years later, "useful mutations did occur in the evolutionary line which produced man, for otherwise, obviously, mankind would not be here."[10]

This begs a huge question.

Of course, genetic change accompanies evolution; but this is no more than a truism.

The question is whether we are seeing cause and effect. Are genes the genesis of all evolution? Or is some other process involved, in which they simply play a part?

The main function of the genetic system, quite clearly, is one of renewal, of maintenance of the *status quo*, of establishing limits to change. Living cells duplicate themselves with near-total fidelity. The degree of error is so tiny that no manmade machine can approach it. There are also built-in constraints. Plants reach a certain size and refuse to grow any larger. Fruit flies refuse to become anything but fruit flies under any circumstances yet devised. The genetic system, as its first priority, conserves, blocks, and stabilizes.

So can the rare mistake that the system makes be held solely responsible for the emergence of new life forms? Can we be sure that the known rate of random gene mutation is sufficient in itself to account for macro-evolution?

For all that the population geneticists unanimously say yes, there are increasing doubts.

CHAPTER THREE

Odds Against

We know almost nothing of the genetic changes that occur in species formation.

—Richard Lewontin,
The Genetic Basis of Evolutionary Change

The primary puzzle is how the genes acquired a meaningful code of information in the first place; and the secondary puzzle, how the code elaborated itself, also in a meaningful way.

The code is at the heart of the prodigious variety of life on Earth today. About 15 million species have already been identified, and there may be as many more again. Creatures range in size from the foot-and-mouth virus some ten millimicrons in diameter to the 100-foot-long blue whale weighing 150 tons. Certain trees are as heavy as 6,000 tons, and others live as many years. The extremes of environment within which bacteria thrive stretches from $-23°C$ in the Antarctic to $+85°C$ in the volcanically heated springs of Yellowstone Park.

Yet all these things, plant and animal life alike, are kept alive by coded messages from the genes that are transmitted by the same method: a ceaseless one-way stream of instructions from DNA via its "messenger" molecule RNA to the proteins that form the chemical basis of life.

There are many other biochemical universals (the term given to properties shared by all living things). All proteins are made up of the same twenty amino acids, in differing proportions. All life is structured from cells of about the same size, which divide and renew themselves in a remarkably similar way. The spiral structure of DNA, with its links consisting of just four nucleotide acids, is also common to everything alive.

These shared properties confirm in the minds of virtually all biologists that everything which lives is derived ultimately from a common source. "In all likelihood, life arose only once," Theodosius Dobzhansky said in 1963.[1]

If so, the mystery of how it arose may never be known. Unique events cannot be investigated satisfactorily by science. If something happened only once, how can we be sure that we are faithfully recreating the conditions in which it happened?

Nevertheless the workings of the genes have become so central to modern evolutionary thinking, and so much depends on their place in the scheme of life, that it is instructive to dwell for a while on what is thought to have taken place.

Beginnings of Life

About 4,600 million years ago (one quarter of the age of the Universe), a hostile, barren Earth was formed from the accumulation of cosmic debris. Between then and the beginning of the Cambrian some 600 million years ago, a number of mountainous steps had to be taken to change the inert dust and gases into such complex living wonders as the sea snails, jellyfish, octopuses, trilobites and so on that abruptly appear in the fossil record.

The first and perhaps the easiest stage was the change from inorganic to organic—from the gases which persumably surrounded Earth at that time (hydrogen, ammonia, methane, etc.) to the simplest amino acids, containing about ten atoms, which are the most basic of the biochemical universals.

Experimentally, Stanley Miller in the United States showed in 1953 that by passing an electrical discharge (in real life, perhaps a bolt of lightning) through the appropriate gases, quite surprisingly large amounts of amino acids were formed. The experiments are acknowledged as a major breakthrough in our understanding of how life got under way, and since then other essential chemicals have been synthesized.

Today, five of the twenty amino acids common to us all still resist attempts to create them artificially under anything like plausible conditions, and critics have pointed to the "oxygen-ultraviolet conundrum" that is still not resolved (panel 7). A summary paper in *Nature* concluded bleakly that the chances of finding significant concentrations of organic chemicals in any prebiotic "soups" so far imagined were vanishingly small: "The physical chemist, guided by the proved principles of chemical thermodynamics and kinetics, cannot offer any encouragement to the biochemist who needs an ocean full of organic compounds."[2]

However, overall the experiments seem to show conclusively

that relatively complicated—although still lifeless—chemicals can arise "spontaneously" from simple ones. There is also the possibility, persuasively argued by the British astronomer Sir Fred Hoyle, that organic compounds exist in outer space, and arrived from there to colonize Earth.[3]

The Biological Treadmill

But the really crucial stage is the one that necessarily followed: the transformation from nonlife to life. Living things are distinguished from nonliving things in a number of ways. Primarily, they organize themselves on a continuing basis. They tend to move themselves toward an increase in order and complexity—they grow, while nonliving things inevitably disintegrate. They have a unique ability to renew themselves after injury—they are able, both at the cellular level and as complete organisms, to reproduce their form. There are other secondary characteristics, such as the ability to dispose of waste, and to respond to stimuli, that have no exact parallel in the nonliving world.

This is true even of the simplest living forms: single-celled bacteria. They represent a quantum evolutionary leap from the lifeless chemicals that came before. Assuming that there was, around four billion years ago, a sea with perhaps a ten percent solution of amino acids, sugars, phosphates, and so on, two prodigious leaps have to take place, *and they have to happen in synchrony.*

The amino acids must link together to form proteins; and the other chemicals must join up to make nucleic acids, including the vital DNA. The seemingly insurmountable obstacle is the way the two reactions are inseparably linked—one can't happen without the other. Proteins depend on DNA for their formation. But DNA cannot form without preexisting protein.

This biological treadmill greatly worries all biologists actively concerned in research into the origin of life. The puzzle was put succinctly: "How, when no life existed, did substances come into being which, today, are absolutely essential to living systems, yet which can only be formed by those systems?"[4]

No one knows the answer. "Which came first?" asked Professor Sidney Fox of Miami University. "Whichever postulate has been considered has seemed to leave an unresolved question."[5]

Nor is it relatively easy chemistry, as in the formation of the

PANEL 7
How Did the First Amino Acids Survive?

It was a Russian biochemist, A. I. Oparin, who in 1936 first suggested how inert chemicals might link together into an organic chain. Although it was impossible to create life from nonlife in our present oxygen-heavy environment, he said (oxygen literally eats up any primitive organic chemical such as an amino acid), this might not have been the case in conditions billions of years ago.

He suggested that there was a "reducing" atmosphere—free of oxygen, and consisting of such gases as methane, ammonia, water and hydrogen. All experiments, including Stanley Miller's, have been based on this hypothesis.

Without oxygen, there is no ozone canopy to protect Earth from the sun's ultraviolet rays. Nowadays, as established by NASA's early space probes, this canopy blankets us between fifteen and thirty miles above Earth's surface, effectively shielding us from certain death.

So with oxygen in the air, the first amino acid would never have got started; without oxygen, it would have been wiped out by cosmic rays.

Imaginative and elaborate solutions have been written to this conundrum. Perhaps the amino acid was formed at the edge of a volcano, and then sank into a lake where it dropped the few meters below the surface necessary to protect it from radiation; perhaps the Earth's waters were covered by a layer of tarlike chemicals which stopped ultraviolet light; perhaps the amino acid was protectively dehydrated or "frozen" in some way on dry rock or clay, waiting for an improvement in the atmosphere.

For every suggestion, there is a seemingly insuperable objection: beneath the surface of the water there would not be enough energy to activate further chemical reactions; water in any case inhibits the growth of more complex molecules; unlike conditions in laboratory experiments, the amino acids and their constituents could not be kept pure and isolated.

In other words, the theoretical chances of getting through even this first and relatively easy stage in the evolution of life are forbidding.

amino acids. Proteins are highly complex biological molecules—
long chains of amino acids strung out like pearls on a necklace,
often twisted and folded in a specific way. Where an amino acid
typically has ten atoms, a protein may have thousands.

Research sponsored by NASA, to enable astronauts to recognize the most rudimentary forms of life, suggested that the
simplest kind of living thing would contain at least 124 proteins
of 400 amino acids each. A genetic code would be functioning,
making sure the organism reproduced true to type.

Now that we have learned to build self-replicating robots, we
can make some calculations of the magnitude of difficulty in
creating a life form even as "simple" as that. Marcel J. E.
Golay, from the viewpoint of an engineer, wrote in *Analytical
Chemistry:*

> Suppose we wanted to build a machine capable of reaching
> into bins for all its parts, and capable of assembling from
> these parts a second machine just like itself. What is the
> minimum amount of structure or information that should be
> built into the first machine? The answer comes out to be of
> the order of 1,500 bits—1,500 choices between alternatives
> which the machine should be able to decide. This answer is
> very suggestive, because 1,500 bits happens to be also of
> the order of magnitude of the amount of structure contained
> in the simplest large protein molecule which, immersed in a
> bath of nutrients, can induce the assembly of those nutrients
> into another large protein molecule like itself, and then
> separate itself from it.[6]

The problem is that a self-replicating robot has been designed
to build copies of itself, whereas the chemical reactions that led
to the first living organism must have happened entirely by
chance. Yet the odds against such a chance occurrence seem
insuperable—by any statistical standards, plain impossible.

Golay calculated the odds of his robot reaching into the bin at
random and sticking all its component bits and pieces together
haphazardly, and then finishing with a perfect copy of itself:
there was only one chance in 10^{450}, he said. (10^{450} is a convenient way of writing 10 with 450 noughts after it. Thus $10^2 = 100$,
$10^3 = 1,000$, and so on.) Frank B. Salisbury in *American Biology
Teacher*, using different calculations, concluded that the odds of
the chance evolution of a medium-sized protein of 300 amino

acids was about one in 10^{600}—a number "completely beyond our comprehension."[7]

Since scientists generally rule out of consideration any event that has less than one chance in 10^{50} of occurring, it is hardly surprising that many have come to believe that new biochemical laws will be needed to solve the problem. However, we know the event did occur, and we even know roughly when.

Earth's First Fossils

Somewhere around 3.5 billion years ago, if geologists have dated things correctly, the first true living organisms had bucked the statistical odds and completed their improbable journey. To the surprise of palaeontologists, who had hardly expected such luck, a number of these incredibly ancient, subtle life forms have been found as micro-fossils in Australia and South Africa. Most of them are apparently identical to modern bacteria and blue-green algae; some are strangers, with minute filaments radiating and branching from the body.

Even the simplest bacterium has about 2,000 genes, each with 1,000 "letters" instructing the organism how to function and reproduce. Here, according to Darwinian theory, was the raw material which led ultimately to the vast library of information in human DNA—the equivalent of 1,700 thousand-page volumes, it is said.[8]

Cellular Complexity

Yet we are absolutely ignorant about how this information got there in the first place, and almost as ignorant about how, at some time during the next billion years, the first true single-celled creatures arose (i.e., cells with a nucleus enclosed by a membrane). Assuming they "grew" from bacteria, and didn't just arrive out of thin air, it is a quantum jump in complexity as statistically improbable as the arrival of a living chemical in the first place. Professor William Thorpe of Cambridge University's zoology department told fellow scientists:

I think it fair to say that all the facile speculations and discussions published during the last ten to fifteen years explaining the mode of origin of life have been shown to be far too simple-minded and to bear very little weight. The problem in fact seems as far from solution as it ever was. The origin of even the simplest cells poses a problem hardly less difficult. The most elementary type of cell constitutes a "mechanism" unimaginably more complex than any machine yet thought up, let alone constructed by man.[9]

The consequences of this ignorance have been far-reaching. For creationist scientists, it confirms their belief that if a designer is needed for a self-replicating robot, then there must surely have been a Great Designer for living organisms—there is no other way, they say, to overcome the statistical odds (panel 8).

For other biologists, seeking an explanation that does not invoke the supernatural, the sheer intractability of the problem has led many to shrug it into the background. Instead of looking at the imponderables of how the code arose, they have studied the minutiae of the code itself.

One ironic result has been an evolutionary theory known as sociobiology, in which the gene becomes scarcely less than godlike in its nature. "If there is an ultimate indivisible fragment it is, by definition, the gene," according to the U.S. biologist George G. Williams, one of the pioneers of this curious doctrine.[10] Richard Dawkins of Oxford University, whose book *The Selfish Gene* helped popularize the idea, argued that

> the fundamental unit of selection, and therefore of self-interest, is not the species, nor the group, nor even, strictly, the individual. It is the gene, the unit of heredity.[11]

In other words, we need not bother about the origin of the genetic system, because the genes *are* life.

DNA is seen as possessing an inbuilt drive to preserve and renew itself, purposefully using the living organism as a way of doing so. The aphorism "a chicken is merely the egg's way of making another chicken" becomes, in sociobiological terms, "a chicken is a device invented by chicken genes to enhance the likelihood of more chicken genes being projected into the future."[12]

In sociobiology, genes rule. Altruism, to philosophers and saints the noblest mode of action—that one man should sacrifice his life for another, or a bird sound an alarm to warn its flock, or worker bees die to protect their queen—is said to be due wholly to a gene. J. B. S. Haldane, one of the founders of the synthetic

PANEL 8
Can Life Form by Chance?

Creationist science literature (see chapter five) makes much of the statistical games to be played with probability theory. Here is Dr. Jean Sloat Morton, who took a Ph.D. in cellular studies at George Washington University, writing in *Impact*, December 1980, number 90 in a series of leaflets issued by the Institute for Creation Research in San Diego, California.

(Be warned: her argument assumes that the only choice is between step-by-step genetic mutations, and instantaneous Divine creation: Darwin vs. God. In chapter seven, other possibilities are raised.)

. . . let us consider a simple protein containing only 100 amino acids. There are 20 different kinds of L-amino acids in proteins, and each can be used repeatedly in chains of 100. Therefore, they could be arranged in 20^{100} or 10^{130} different ways. Even if a hundred million billion (10^{17}) of these combinations could function for a given purpose, there is only one chance in 10^{113} of getting one of these required amino acid sequences in a small protein consisting of 100 amino acids.

By comparison, Sir Arthur Eddington has estimated there are no more than 10^{80} (or $3,145 \times 10^{79}$) particles in the universe. If we assume that the universe is 30 billion years old (or 10^{18} seconds), and that each particle can react at the exaggerated rate of one trillion (10^{12}) times per second, then the total number of events that can occur within the time and matter of our universe is $10^{80} \times 10^{12} \times 10^{18} = 10^{110}$. Even by most generous estimates, therefore, there is not enough time or matter in our universe to "guarantee" production of even one small protein with relative specificity.

If probabilities involving two or more independent events are desired, they can be found by multiplying together the probabilities of each event. Consider the 10 enzymes of the glycolytic pathway. If each of these were a small protein having 100 amino acid residues with some flexibility and a probability of 1 in 10^{113} or 10^{-113}, the probability for arranging the amino acids for the 10 enzymes would be: $P = 10^{-1,130}$ or 1 in $10^{1,130}$.

And 1 in $10^{1,130}$ is only the odds against producing

the 10 glycolytic enzymes by chance. It is estimated that the human body contains 25,000 enzymes. If each of these were only a small enzyme consisting of 100 amino acids with a probability of 1 in 10^{-113}, the probability of getting all 25,000 would be $(10^{-113})^{25,000}$, which is 1 chance in $10^{2,825,000}$. The actual probability of arranging the amino acids of the 25,000 enzymes will be much slimmer than our calculations indicate, because most enzymes are far more complex than our illustrative enzyme of 100 amino acids.

Mathematicians usually consider 1 chance in 10^{50} as negligible. In other words, when the exponent is higher than 50, the chances are so slim for such an event ever occurring, that it is considered impossible. In our calculations, 10^{-110} was considered the total number of events that could occur within the time and matter of our universe. The chances for producing a simple enzyme-protein having 100 amino acid residues was 1 in 10^{113}, the probability for 25,000 enzymes occurring by chance alone was 1 in $10^{2,825,000}$. It is preposterous to think that even one simple enzyme-protein could occur by chance alone, much less the 10 in glycolysis or the 25,000 in the human body.

... the tight fit among complex and interdependent steps—especially the way some reactions take on meaning only in terms of reactions that occur much later in the sequence—seems to point clearly to creation with a teleological purpose, by an Intelligence and Power far greater than man's.[28]

---

theory, once remarked that he was prepared to lay down his life to save two brothers or eight cousins—the idea being that his "altruism" gene would persuade him to make the supreme sacrifice if, by so doing, the gene would ultimately increase in frequency through his descendants in the population.[13] It was the same gene that Mary Williams suggested would act to maximize the numbers of grandchildren and possible great-great-grandchildren too.[14]

Within this "absurd and degenerate concept," as it is described by one critic, "entities are freely invented and endowed with whatever properties are required to 'explain' biological phenomena."[15] Another wrote that in sociobiology no attempt was made to discover what laws might lie behind the formation of the genetic code. Instead, "genes are invented whenever there is something to explain or explain away. There is nothing beyond the reach of natural selection, given the demon-like powers of suitable genes which are capable not only of local effects, but of action at a distance as well."[16]

Weismann's Barrier

Even neo-Darwinists who do not take such an extreme view as the sociobiologists still wholeheartedly accept a principle put forward in pre-gene days by the nineteenth-century German biologist August Weismann. Weismann's doctrine, or Weismann's

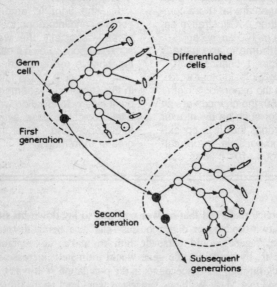

Germ cells, according to Weismann's doctrine, are passed unchanged from generation to generation. It is why species remain intact, and why you can amputate a limb without harm to the offspring—the somatic cells (body cells, or "differentiated" cells) are kept by a biological barrier from the germ cells. Recent experiments suggest this may not always be the case.

barrier, says that there are special reproductive cells which are responsible for passing on parental characteristics to the offspring (known as *germ* cells, as distinct from *somatic* cells in the rest of our body—the soma). Germ cells become isolated very early in the growth of the embryo, and remain unaffected by what happens in the rest of the body during the remainder of its life. It explains, for instance, why you can chop off wings from adult *Drosophila*, but the young fruit flies are still born normal. The wing amputation has affected the somatic cells, but not the germ cells.

Weismann saw in this a "central directing agency," much as neo-Darwinists see genes playing the same role; its primary job was to secure "continuity of the germ-plasm." The essence of the belief, updated by the geneticists, is that nothing can happen to DNA from without. It is irretrievably enclosed in a protective capsule, constantly emitting coded messages that instruct the organism how to act. Occasionally a random mutation takes place, the message changes, and the organism acts accordingly. The organism can never answer back. Come heat, come cold, come rain, come sun, we are at the mercy of our genes.

Now this sometimes bothers even those who agree that, as a general rule, Weismann's separation of germ cells from somatic cells is valid. If nothing can come from without—if the barrier is total and permanent, and can never, without exception, be penetrated—biology is forced into a precarious assumption: the first living creature must have had within itself the entire genetic potential to grow into—to *create*—every one of the trillions of plants and creatures that have lived since. DNA becomes not only the library and transmitter of information, but its sole author. "To attribute such a power to a single substance, however complicated and exceptional its molecular structure may be, is in my view aberrant," wrote Professor Pierre-Paul Grassé, the renowned French biologist.[17]

Molecular Darwinism

Biologists, it seems fair to conclude, are unanimously ignorant about the origin of the genetic code. So what about the mechanism that allowed it to elaborate itself?

On the credit side, after more than half a century of molecular biology, there is now a secure theory of how heredity works. We can also accept that there is a Darwinian selection process among microorganisms (bacteria, viruses).

Among the experimental triumphs has been that by James Watson and Francis Crick in 1953 when they unraveled the crystalline structure of the DNA molecule, and thereby demonstrated biology's "central dogma" that coded information passes only from DNA to protein, never the other way around. In the general consensus of scientists, it constituted the greatest intellectual achievement of modern science, for it showed at a molecular level the method by which characteristics are passed on from parents to offspring.

Scarcely less important has been the work of Fredrick Sanger of Cambridge University, winner of two Nobel prizes, the first person to map the entire structure of a gene. He has shown that most mutations are single changes in a DNA sequence of "letters," and that these mutations (called "point mutations") cause observable alterations in protein. The altered instructions from the gene create a different kind of protein. There is, therefore, proven cause and effect at a molecular level: mutations bring about chemical changes.

Earlier, in 1943, came a key experiment in the United States by S. E. Luria and M. Delbruck. It is relied upon very heavily by neo-Darwinists, for it showed that random mutations in bacteria enable them to resist virus attacks and thus that microevolution takes place at the molecular level. According to the biochemist Jeffrey Pollard of Queen Elizabeth College, University of London, "this conclusively demonstrated that variation occurred before the environmental selection, and put Darwinism on a very firm molecular basis."[18]

Other experimental evidence in support of a genetic basis for evolutionary theory has come from the discovery that whole batteries of genes are switched "on" and "off." A single mutation (or a changed instruction) here presumably has a much bigger effect than the "single letter" change demonstrated by Sanger—a mechanism, perhaps, for speeding up variation, and providing a basis for macro-evolution.[19]

Biological Ignorance

But already we have to use the words "presumably" and "perhaps."

As Jeffrey Pollard continued: "By necessity, all the above experiments were performed on micro-organisms, and you can't argue with neo-Darwinism at this level. However, the big evolutionary questions are about multi-cellular organisms—Darwin's origin of species, which he never properly solved. It is very much an open question whether the mechanisms that we see in a bacterial germline are the only ones happening in the production of new species."

Indeed, each new experimental finding tends to compound our ignorance about the basic processes of DNA, as much as elucidate them. For instance, there is no obvious advantage in having a large quantity of DNA, as you might expect. True, you and I have about 100 times as much DNA as bacteria. But there is no stately progressive evolutionary increase to be seen, no simple correlation between complexity of form and amount of DNA—a salamander, assuredly less complex than we humans, has twenty times as much as us. The list of similar puzzles is almost endless (panel 9); here are a few that have been thrown up by recent research:

- DNA in chimpanzees differs from our own by less than one percent. This worries geneticists, who would have expected a far bigger difference to account for the enormous dissimilarity between the two species.
- There is no simple one-for-one correlation between genes and outward appearance—you can get more than one protein from the same DNA sequence.
- We know that genes are turned on and off—but how?
- No gene for altruism has been found—indeed, nor has a "behavior" gene of any kind.
- And perhaps the most profound riddle of all: how do cells "decide" what their future job in the organism is to be, when each has the same set of genetic instructions? Early in embryonic development, cells start differentiating and eventually become quite different substances—some become skin cells, some tissue for the stomach or heart, some liver, and so on. Yet each cell, so far as we know (including germ cells) contains identical DNA.

Roy J. Britten of the California Institute of Technology, writing in the *Encyclopaedia of Ignorance*, concluded that from what had been discovered so far, mutations did not happen often enough to account for evolutionary change; the highly organized

PANEL 9
How Do Genes Mutate?

Roy J. Britten, in the *Encyclopaedia of Ignorance,* a book in which experts describe the limits of scientific knowledge, wrote at some length on "The Sources of Variation in Evolution" before summing up in a final paragraph: "In conclusion I wish to emphasize that while ingenious speculations can be made about the gene regulation system and its relationship to the source of variation in evolution, our ignorance of the actual sources of variation is just as abysmal as our ignorance of the formal properties of the regulatory system that is evolving."

Earlier, he listed some of the speculations and question marks that surrounded the type of mutations that might be inherited and lead to new forms of life:

Are they ordinary mutations (base substitutions in genes coding for proteins) or are they more complex changes in the DNA? Are visible chromosomal rearrangements more important? Are there many rearrangements of short DNA sequences? What about changes in the "system of gene regulation"? Changes in the regulation of genetic activity would be very important to form and development. Rearrangements could add and subtract given genes from a coordinately expressed set of genes. Changes on the level of the regulatory system itself could have immense potential power. Large blocks of genes could be affected. Whole branches of the regulatory system could be turned on or off that contain extensive information capable of specifying an organ or organ part. These "turned-off" sets of regulatory system information could be preserved for extensive periods. In this way fossil organs, organ systems or system parts could be preserved in unexpressed or occasionally expressed form as preserved regulatory patterns. The re-expression of combinations of such patterns could produce novel structures with unexpected potentiality. Interesting suggestions, but the problem before us is the development of some direct evidence.[29]

PANEL 10
What Do Mutations Achieve?

The first major objection to genes being the sole and sufficient driving force for evolution is that practically every mutation is obviously harmful, and puts the organism at a disadvantage rather than an advantage. Two of the most powerful causes of mutation are mustard gas and x-rays. A moment's reflection on the horror of Hiroshima children born with deformed limbs and bodies, or blood disorders condemning them to premature deaths, is enough to show that they were unlikely candidates, to say the least, to win the struggle for existence in a life-game where survival of the fittest is the governing rule.

C. P. Martin, of McGill University's biology department, once compared the way that x-rays work on the body's metabolism to a person being kicked and beaten in a random, mindless manner—it was impossible for x-rays to cause anything but damage, he said, even if the damage was subsequently somewhat repaired.

"It is quite possible that the violent knocking about might dislocate a man's shoulder, and that continued knocking about might actually reduce a previous dislocation," he continued. "No sane person would cite such a case as this to prove that the result of knocking a man about are not injuries; nor would anyone refer to the result as evidence that knocking a man about can produce an improvement over the normal man."[30]

So he concluded that a mutation was a pathological process that had nothing to do with evolution, and that the rare occasions when one proved helpful had been isolated flukes that did not constitute a general evolutionary mechanism.

Even Theodosius Dobzhansky, while coming to the opposite conclusion, admitted the problem. "A majority of mutations, both those arising in laboratories and those stored in natural populations, produce deteriorations of viability, hereditary disease, and monstrosities. Such changes, it would seem, can hardly serve as evolutionary building blocks."[31]

The genetic makeup of chimpanzees and humans is more than 99 percent identical, according to molecular biologists. It seems to mean we had a common ancestor less than five or six million years ago (see chapter eight), and is a puzzle to geneticists who would have expected a bigger difference to account for the anatomical dissimilarity. Victorian illustrators, however, were quick to discover the comic possibilities in "Mr. Darwin's monkey theory."

state of DNA was unexplained; and the gene regulatory system operates in an unknown way.

"We are ignorant of the sources of variation in evolution."[20]

In the absence of a theory of how the genetic code originated, and amid doubts as to how it elaborated, neo-Darwinists have come to rely on the sheer size of the information potential that it contains. The importance of random beneficial mutations has been downgraded (panel 10), and instead, the mixing of genes that happens during sexual reproduction is emphasized. Ernst Mayr has said that out of every 100 new genes, only one would come from mutation, and ninety-nine from gene flow (sexual mixing in populations).

Again, nobody knows how or precisely when this first happened (it is part of the Pre-Cambrian jump from bacteria, whose sex life is infrequent and inefficient, to all higher creatures built on cells with a nucleus and a membrane). But once organisms started sharing DNA from the mother and the father, there began a potentially prolific source of hereditary variation.

According to neo-Darwinists, the immense quantity of DNA that resulted easily made possible the evolution of anything that has ever lived on Earth.[21] It is because certain characteristics of this vast genetic pool can be measured statistically—the rate of mutation, of gene flow, and so on—that the discipline of population genetics has taken such a firm hold. The mind-numbing pages of trivial applications of binomial theorem, and arcane debate about what proportion of mutations are harmful, beneficial, or neutral (panel 11), have seemed to give neo-Darwinism mathematical validity.

Ernst Mayr, the high priest of this kind of thinking, is confident that "all evolution is due to the accumulation of small genetic changes, guided by natural selection," that the origin of species is "nothing but the extrapolation and magnification of the events that take place within populations and species," and that attacks against this "are either based on ignorance or are motivated ideologically."[22]

PANEL 11
Are "Neutral" Mutations Behind Evolution?

The theory of neutral mutations has been developed in a number of ways. Expressed simply, it says that many mutations are neither beneficial nor harmful, and are therefore not selected for or against by natural selection. In the normal course of events they will drift out of the population at a rate which can be calculated.

Thus the chances are about two to one that such a mutation will survive into the next generation, six to one that it will have disappeared by the tenth generation, about fifty to one that it will be gone by the hundredth generation, and there is only one chance in a thousand that it will persist for a thousand generations.[32]

However, probability theory also teaches us that occasionally, at random, a number of such mutations will become "fixed" in the population despite the odds against; and also that this is much more likely to happen in a very small population.

This fits well with Ernst Mayr's notion of evolution starting with small, peripheral, isolated groups of plants or creatures. "Favoring in every generation certain individuals owing to some properties they have, natural selection chooses automatically all their other genes, including many that are near-neutral or even slightly deleterious. When such selection happens in very small gene pools, rather pronounced departures from the optimal genotype sometimes survive, owing to errors of sampling."[33] (*Genotype* is defined as the genetic makeup of a single organism.)

Genetic drift has been shown to be responsible for certain characteristics in fruit flies that do not affect survival one way or the other: bristle number, wing color variations, and so on. More controversially, the theory has been used to "solve" the problem discussed in the next chapter: the apparent need for the simultaneous occurrence of a number of unrelated advantageous mutations in a single animal in one generation.

According to this hypothesis, previously neutral mutations, lying dormant in the animal, can be "activated" by a new mutation so that suddenly they confer a selective advantage. The classic example of a giraffe's neck might be explained by assuming that over a period of thousands of generations a series of mutations accumulated stipulating stronger bronchial arches, greater musculature, and a bigger heart. Although for the time being there is no "need" for them, they will, with a degree of probability, become fixed in the population. Then, if a mutation occurs which causes a long neck, the support systems to allow the neck to be developed are ready to be activated, giving to the external eye the apparent appearance of several unrelated mutations occurring in one animal in one generation.

A great deal of mathematical effort has been expended to plot frequencies of occurrence of neutral mutations, probabilities of coincident occurrence, etc. The problem with the theory, as with population genetics as a whole, is that you can play the numbers game to "prove" almost anything you like, as neatly summarized by Colin Patterson of the British Museum of Natural History:

> Darwinian evolution, by natural selection, predicts that organisms are as they are because all their genes have been and are being subjected to selection, those that reduce the organism's success being eliminated, and those that enhance it being favoured. This is scientific theory, for these predictions can be tested. "Non-Darwinian" or random evolution predicts that some features of organisms are non-adaptive, having neutral or slightly negative survival value, and that the genes controlling such features are fluctuating randomly in the population, or have been fixed because at some time in the past the population went through a bottleneck, when it was greatly reduced. When these two theories are combined, as a general explanation of evolutionary change, that general theory is no longer testable. Take natural selection: no matter how many cases fail to yield to a natural selection analysis, the theory is not threatened, for it can always be said that these failures of selection theory are explained by genetic drift. And no matter how many supposed examples of genetic drift are shown to be due, after all, to natural selection, the neutral theory is not threatened, for it never pretended to explain all evolution.[34]

Scientific Doubts

Yet doubts remain, even among geneticists and biologists. Professor Sir Ernst Chain, who won a Nobel prize for research into the curative properties of penicillin, said in 1970:

> To postulate that the development and survival of the fittest is entirely a consequence of chance mutations seems to me a hypothesis based on no evidence and irreconcilable with the facts. These classical evolutionary theories are a gross over-simplification of an immensely complex and intricate mass of facts, and it amazes me that they are swallowed so uncritically and readily, and for such a long time, by so many scientists without a murmur of protest.[23]

In 1966 there was an inconclusive and often ill-tempered two-day symposium at the Wistar Institute of Anatomy and Biology in the University of Pennsylvania entitled ''Mathematical Challenges to the Neo-Darwinian Interpretation of Evolution.'' Here it became clear that doubts among biologists were doubled and redoubled by physicists, mathematicians and engineers, some of whom were openly incredulous at the lack of a testable scientific basis for evolutionary theory. (Few biologists expressed any uncertainty, on this occasion, about natural selection being the supreme explanatory law, prompting a delegate from the other side to remark: ''If I wanted to be nasty to the evolutionists, I would say that they are surer of themselves than we nuclear physicists are—and that's quite a lot.''[24])

Computer scientists, especially, were baffled as to how random mutations alone could possibly enrich the library of genetic information. A mutation, they repeatedly pointed out, is a mistake—the equivalent of a copying error. And how could copying mistakes build up into a new body of complicated and ordered information? Murray Eden, Professor of Engineering at Massachusetts Institute of Technology, said that in plain language what the biologists were proposing went as follows:

> The chance of emergence of man is like the probability of typing at random a meaningful library of one thousand volumes using the following procedure: Begin with a mean-

ingful phrase, retype it with a few mistakes, make it longer by adding letters; then examine the result to see if the new phrase is meaningful. Repeat this process until the library is complete.[25]

He concluded this was so implausible that "an adequate scientific theory of evolution must await the discovery and elucidation of new natural laws—physical, physico-chemical and biological."

Marcel P. Schutzenberger, a computer scientist from the University of Paris, agreed that spontaneous improvement and enlargement of the code through mutations and natural selection was "not conceivable."

> In fact, if we try to simulate such a situation by making changes randomly at the typographic level (by letter or by blocks, the size of the unit does not really matter) on computer programmes, we find that we have no chance (that is, less than one chance in $10^{1,000}$) even to see what the modified programme would compute: it just jams.

And he summed up: "We believe that there is a considerable gap in the neo-Darwinian theory of evolution, and we believe this gap to be of such a nature that it cannot be bridged within the current conception of biology."[26]

Repeatedly, the impasse that Darwinians reach is one with a sign saying there is no useful road ahead without a map showing the overall pattern of evolution; random mutations—at least, sequential mutations of a one-after-one Darwinian kind—lead to yet more dead ends.

Nowhere is this more true than in the biological enigma we come to next: what Darwin called "organs of perfection." To quote Richard Lewontin of Harvard again, who provided the keynote admission at the beginning of this chapter that almost nothing is known about the genetic changes involved in species formation, many organisms "appear to have been carefully and artfully designed."[27]

It is, he says, both a challenge to Darwinism and "the chief evidence of a Supreme Designer."

CHAPTER FOUR

Biological Oddities

> If it could be demonstrated that any complex organ existed which could not possibly have been formed by numerous successive slight modifications, my theory would absolutely break down.
>
> —Charles Darwin,
> *The Origin of Species*

There were a number of nature's wonders that gave Darwin the shudders. A feather in a peacock's tail was one. "Small trifling particulars of structure often make me very uncomfortable," he confessed not long after publication of *Origin*.

But most worrying of all was that marvel of construction, the human eye.

For the eye to work the following minimum perfectly coordinated steps have to take place (there are many others happening simultaneously, but even a grossly simplified description is enough to point up the problems for Darwinian theory). The eye must be clean and moist, maintained in this state by the interaction of the tear gland and movable eyelids, whose eyelashes also act as a crude filter against the sun. The light then passes through a small transparent section of the protective outer coating (the *cornea*), and continues via a self-adjusting aperture (the *pupil*), and a similarly automatic *lens* which focuses it on the back of the *retina*. Here 130 million light-sensitive rods and cones cause photochemical reactions which transform the light into electrical impulses. Some 1,000 million of these are transmitted every second, by means that are not properly understood, to a brain which then takes appropriate action.

Now it is quite evident that if the slightest thing goes wrong *en route*—if the cornea is fuzzy, or the pupil fails to dilate, or the lens becomes opaque, or the focusing goes wrong—then a recognizable image is not formed. The eye either functions as a

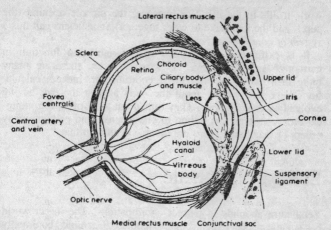

*Section through a human eye. All its parts must work together.
But could this synchronism have evolved by chance?*

whole, or not at all. So how did it come to evolve by slow,
steady, infinitesimally small Darwinian improvements? Is it really possible that thousands upon thousands of lucky chance
mutations happened coincidentally so that the lens and the retina,
which cannot work without each other, evolved in synchrony?
What survival value can there be in an eye that doesn't see?

Small wonder that it troubled Darwin. "To this day the eye
makes me shudder," he wrote to his botanist friend Asa Gray in
February 1860.

The eye is, in fact, merely an extreme example of a large
number of *evolutionary novelties,* as they have come to be
termed—structures that, logically, have either to be perfect, or
perfectly useless. Darwin himself called them "organs of extreme
perfection and complication." With each of them, the difficulty
for Darwinians is twofold. Theory demands that successive steps
of a gradually improving nature build toward a final product
perfectly adapted to its environment. But many of the proposed
intermediate steps seem impractical or even harmful. What use
would be half a jaw? Or half a lung? Natural selection would
surely eliminate creatures with such oddities, not preserve them.

Secondly, simultaneous advantageous mutations seemingly have
to take place. Otherwise the organ, even half-formed, would not

work at all. In the eye, for instance, the pinhole opening (the pupil) and the lens have to work together. Statisticians call this a *system of coordinated variables*.

It is extremely difficult for two variables to function in harmony—and in the eye, as we have seen, there are many more than two. The problem is coordination. Indeed, calculations have been made about the odds against the eye having evolved by chance alone. They turned out to be of an astronomical order—at least ten billion to one against, and perhaps many orders of magnitude more improbable even than that.[1]

Bombardier Beetle

Yet nature abounds in perfect coordinations. The insect world has some especially spectacular examples. *Brachinus*, commonly known as the Bombardier Beetle, squirts a lethal mixture of hydroquinone and hydrogen peroxide into the face of its enemy. These two chemicals, when mixed together, literally explode. So in order to store them inside its body, the Bombardier Beetle has evolved a chemical inhibitor to make them harmless. At the moment the beetle squirts the liquid out of its tail, an anti-inhibitor is added to make the mixture explosive once again. The chain of events that could have led to the evolution of such a complex, coordinated and subtle process is beyond biological explanation on a simple step-by-step basis. The slightest alteration in the chemical balance would result immediately in a race of exploded beetles.

The problem of evolutionary novelties is quite widely accepted among biologists (panel 12). In every case, the difficulty is compounded by the lack of fossil evidence. The first time that the plant, creature, or organ appears, it is in its finished state, so to speak. Among the most intractable examples out of many are the following.

Whales

Nobody knows the origin of whales, although it is presumed their ancestors were some kind of primitive hoofed mammals which moved from the land to the sea, and that during this period there "must have been" an amphibious stage. That they are mammals is beyond doubt, for they are warm-blooded, air

PANEL 12
Inexplicable Evolutionary Novelties

The problem of creatures, plants, or organs appearing suddenly in the evolutionary record in a completely new form is widely admitted in academic circles. "The argument still rages," wrote Stephen Gould in a review of Darwin's own treatment of the subject. Ernst Mayr agreed that biologists "occasionally find what appears to be an entirely new structure . . . the bird feather, the ear bones of mammals, the swim bladder of fish, the wings of insects, and the sting of wasps, ants and bees."

Richard B. Goldschmidt, a refugee from Hitler's Germany who spent the remainder of his life as a controversial biologist at UCAL at Berkeley, believed strongly that macroevolution must have happened in mutational leaps, and coined the phrase "hopeful monster" to describe this.[17] He challenged Darwinians to provide explanations for seventeen features which he claimed could not have evolved on a step-by-step basis. They included:

Hair in mammals
Feathers in birds
Segmentation of arthropods and vertebrates
Transformation of gill arches
Teeth
Shells of molluscs
Blood circulation
Poison apparatus of snakes
Whalebone
Compound eyes

The challenge was never comprehensively taken up, although as the text in this chapter shows, there are attempted explanations for some of them.

breathing, give birth to their young alive, and suckle each one with up to one and a half tons of milk a day.

The first creatures classified as whales, *Archaeoceti*, are found in the fossil record roaming the high seas only five million years after the most famous crisis in the history of Earth, the extinction of dinosaurs around sixty-five million years ago. Modern whales arrived about five million years after that.

The problem for Darwinians is in trying to find an explanation for the immense number of adaptions and mutations needed to change a small and primitive earthbound mammal, living alongside and dominated by dinosaurs, into a huge animal with a body

PANEL 13
How Did the Whale Get Its Tail?

One of the principal problems for Darwinians in whale evolution is constructing a pattern of events for the whale's tail to emerge in small, naturally selected steps. The point is that the tail moves up and down, whereas in a land mammal it moves from side to side. This may sound a relatively small difference, but anatomically it is not. It means that somehow the whale's ancestor had to get rid of its pelvis.

Now this cannot be done just because the animal would prefer things this way. However much it wished that it could move its rear quarters up and down more vigorously so that it could swim faster to catch more fish, it could improve only up to a certain point; after this, the pelvic bones would prevent further movement.

According to Michael Pitman, a young Cambridge University biologist who has made a study of the problem, "every down-ward movement of such a tail would crush the reproductive opening of the creature against the back of the pelvis, causing pain and harm." Taken to the extreme, there would come a point where the pelvis would be fractured by the action of the tail, thereby making survival impossible. Natural selection would work against, not for, such a change.

So for the up-down action in whales to emerge, there *simultaneously* had to be random genetic changes that diminished the pelvis while allowing the tail to grow larger. Apart from the stupefyingly long odds against such a chain of events happening by chance, Pitman has concluded that there is a further anatomical objection. At a certain point in the supposed transitionary period, the hip bone would have been "too small to support the hind legs and yet too large to permit the musculature necessary to move the great tail of the whale."[18]

uniquely shaped so as to be able to swim deep in the oceans, a vast environment previously unknown to mammals.

Notable complexities in whale evolution concern the eye, subtly changed so that light rays through the sea water are brought to focus on the retina; the skin, which has a curious outer surface helping to streamline the flow of water; the replace-

ment of sweat glands by a thick layer of blubber fat to control
the body temperature; the superb hearing system; the way in
which a female whale suckles her young under water without
them drowning; and the plates of *baleen* which hang like curtains,
instead of teeth, from the roof of the mouth of whalebone
whales, acting as perfectly designed sieves for the tiny crustaceans which form their food (panel 13).

All this has to evolve in at most five to ten million years—
about the same time as the relatively trivial evolution of the first
upright walking apes into ourselves.

Mammalian Ear

There are two problems here: the early evolution of the ear, and
its extraordinary sophistication.

As we saw in the first chapter, one of the chief distinctions
between reptiles and mammals is that the former have a single
earbone (the *stapes*) and at least four bones in the lower jaw,
whereas mammals have two extra earbones, but only one jawbone. The evolutionary assumption is that some of the reptile's
jawbones became embedded in the middle ear and finally
transformed into the *malleus* (hammer) and *incus* (anvil) bones.
The lack of evidence that this is, in fact, what happened (there

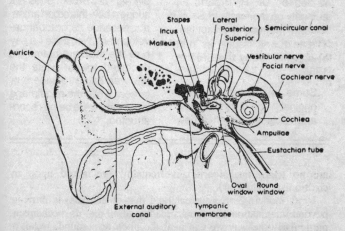

Section through human ear—"intricate beyond imagination."

PANEL 14
The Chewing/Hearing Conundrum

The imagined step-by-step process leading up to the appearance of the mammal ear is, like the human eye, a favorite target in literature ridiculing the whole idea of evolution. This passage is by Douglas Dewar, a Fellow of the Zoological Society of London who became disenchanted with Darwinian theory and helped found the Evolution Protest Movement in Britain. During the ten years before his death in 1957, he was its president. Here he "translates" into plain English an apparently plausible scientific account of the reptile/mammal jaw transformation written by Dr. R. Broom, an authority on the fossils of South African mammallike reptiles. The text was issued as part of a pamphlet by the EPM in 1965.

Some reptile scrapped the original hinge of its lower jaw and replaced it with a new one attached to another part of the skull. Then five of the bones on each side of the lower jaw broke away from the biggest bone. The jaw bone to which the hinge was originally attached, after being set free, forced its way into the middle part of the ear, dragging with it three of the lower jaw bones, which, with the quadrate and the reptilian middle-ear bone, formed themselves into a completely new outfit. While all this was going on, the organ of Corti, peculiar to mammals and their essential organ of hearing, developed in the middle ear. Dr. Broom does not suggest how this organ arose, nor describe its gradual development. Nor does he say how the incipient mammals contrived to eat while the jaw was being rehinged, or to hear while the middle and inner ears were being reconstructed.[19]

are no transitional fossils) is frequently remarked upon by creationists (panel 14).

In its modern form, our ear (and that of animals) is intricate beyond imagination. Textbooks readily admit that its mechanism, particularly the way we can hear the pitch of a sound, is not at all well understood. The incomprehension stems from the

PANEL 15
Do We Hear Well Enough?

Darwin wrote: "Natural selection tends only to make each organic being as perfect as, or slightly more perfect than, the other inhabitants of the same country with which it has to struggle for existence."

The trouble with the mammal ear is that, in terms of natural selection, it has nothing of enough significance to justify its enormously complex system having emerged by natural selection. Amphibians, reptiles, and birds, all of which have only one earbone, can perceive pitch and volume at least as well as mammals, and in some cases better.

The sole possible advantage is that mammals can hear to some extent stereophonically, while it is thought creatures with single earbones cannot do this quite so well. However, barn owls (and probably others as well) can locate prey in the dark by receiving sound stereophonically. The ears are slightly asymmetrically placed on the head. The sound emitted by the prey is received split seconds apart in each ear. The brain decodes the information and works out coordinates. In the case of mammals, stereophony happens because our brains receive signals from both the outer and the inner ear, and the fractional delay in the sound impulses may enable us to estimate how far away a sound is coming from. In survival value, this might confer a minimal advantage in, for instance, spotting prey or escaping predators.

But even if this ability were proved (for doubts still remain), it is hard to see how the transitional forms leading up to it could have made the ear, in Darwin's words "slightly more perfect." The stereophonic effect can only work when the inner and outer ears have been fully displaced.

complexity. The organ of Corti alone, a spiraling 3 mm diameter ridge of cells in the inner ear that seems to play a crucial part in the way we hear pitch and direction of sound, contains some 20,000 rods and more than 30,000 nerve endings.

Yet within this complexity lies a further paradox (panel 15). Although nothing remotely as complicated can be found in the ear of reptiles, living or extinct, it is far from certain that we hear significantly better than they do. So where is the special advantage that, according to theory, would be naturally selected?

Snakes

No zoologist doubts that snakes evolved from reptiles. The transformation of a four-legged into a legless reptile is said to have happened in several distinct varieties. Some snakes—pythons and boas—still have what seem to be relics of their hip bones inside them, and bulges on the outside where probably the legs once grew.

The problem comes in explaining how the snake developed, through a very large number of chance mutations, its unique method of locomotion, which is distinctly unreptilian. It demands an increase in the number of certain vertebrae. Without this crucial variation, snakes could not wriggle about and move in the way they do. Yet it apparently happened in several independent groups of snakes in widely differing parts of the Earth and in different periods.

Similarly, the snake's eating apparatus requires a large number of independent functions to evolve simultaneously: for instance, jawbone modifications, reshaping of the teeth, appearance of an extra row of teeth, special protection for the brain, modifications of the glottis, and appearance of powerful digestive juices. There is no trace of any transitional stages in the fossil record; nor has anyone ever seen, in the broods of four-legged lizards, a mutated monstrosity which might give a clue as to how a reptile can become both legless and perfectly adapted to a new way of locomotion and life.

Origin of Flight

One of the more remarkable facts of evolution is that the ability to fly occurred independently and separately four times among different kinds of creatures.

1. *Insects* Winged insects appeared suddenly and plentifully alongside wingless insects in the Carboniferous period, some 300 million years ago. Before that there are no fossil insects at all, flying or otherwise, so any suggestion as to how they evolved the capacity of flight is sheer guesswork.

2. *Dinosaurs* Flying pterosaurs (e.g., pterodactyls and pteranodons) were abundant from about 180 million years ago until they were wiped out in the general extinction of dinosaurs sixty-five million years ago. Again, the first known fossils show them fully capable of flight, and although the earliest ones were rather less specialized for flight than the later ones, there is absolutely no sign of earlier intermediate stages.

3. *Birds* As described in the first chapter, there is no agreed origin of birds. They arrived plentifully in the fossil record about sixty million years ago, simultaneously equipped with feathers, hollow bones, a new digestive system, air sacs, and a number of other novelties that distinguished them from their reptilian ancestors. Academically, the debate has centered on whether dinosaurs, pterosaurs and birds all had a common ancestor, or whether some early dinosaurs were transformed into birds.

4. *Bats* The last group of creatures to gain the capacity of flight was the bats, fifty million years ago. Again there is no clue as to what transitional stages there may have been. It is presumed they evolved from some earlier, insect-eating, shrewlike animals that climbed trees. Today, there are a few species of animals that can glide from branches of trees. But they are anatomically unlike bats in almost every other respect, and nobody regards them as providing a model for the origin of bat flight.

In particular, it is difficult to imagine intermediate stages for bats. A half-developed forelimb would mean the creature could neither fly nor walk properly. Also, a bat's pelvic girdle has rotated 180° compared with other mammals, enabling it to fly; what possible selective advantage could there be in a 90° turn? The milk teeth of baby bats are turned inwards so that they can cling on to the mother's hair as she hangs high from the ground. If this faculty had not appeared fully operable, would it not have been fatal both for individual baby bats and thus for the entire species?

Looking for explanations of evolutionary novelties, huge in number as they are, it is perhaps not surprising to find that textbooks in general are reluctant to explore these tricky questions. Sir Julian Huxley's classic work *Evolution: The Modern Synthesis*, first published in 1940 and still today, after many editions, a key statement of neo-Darwinian theory, does not mention the evolution of the eye at all. Nor does Paul Amos Moody's *Introduction*

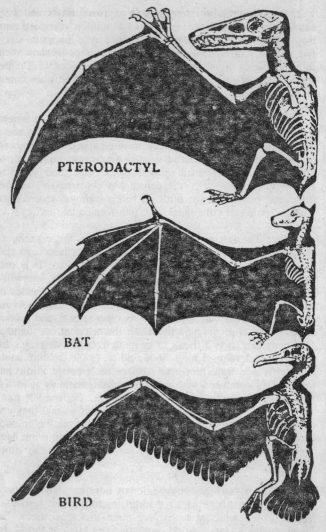

PTERODACTYL

BAT

BIRD

Dinosaurs, bats and birds each evolved the ability to fly by different methods and at different times in Earth's history. None of the transitional stages are preserved in the fossil record.

to Evolution, one of the most widely approved textbooks of the last twenty-five years; nor Grizmek's *Encyclopaedia of Evolution;* nor the *Encyclopaedia Britannica*.

Colin Patterson, in his recent book *Evolution*, published by the British Museum of Natural History, also steers clear of the eye problem, except to admit that a lens in the eye is no use unless it works, and that a distorting lens might be worse than no lens at all.

(He also raises the question of another evolutionary problem: "How can the segmentation of an animal like an earthworm or a centipede arise bit by bit?" Again he drops the subject abruptly apart from confessing: "an animal is either segmented or it is not."[2])

Darwin worried about the eye for four and a half pages in *Origin*, and in the end took comfort in noting the large number of different kinds of eyes that had emerged in various creatures, living and extinct, ranging from the primitive to the near perfect. This line has been followed by his disciples ever since (panel 16), but is open to severe doubt. Eyes from different species can be arranged neatly on the printed page in a graded series, but none of them necessarily has anything to do with how the human eye evolved. It shows, of course, that nature has taken several different paths in creating the ability to see, but this in a way merely aggravates the problem, for the process of evolution in *none* of them is clear.

The abracadabra approach used to explain away the fossil gaps is equally in evidence when it comes to evolutionary novelties. With a wave of the wand, difficulties and complexities are minimized. Thus the truly vast anatomical steeplechase that has to be run to turn a small hoofed animal (presumably) into a whale becomes:

> The major differences between the whales and the early mammals are attributable to adaptions for the swimming life. The forelimbs have become paddles. The rear limbs have been lost altogether, though there are a few small bones buried deep in the whale's body to prove that the whale's ancestors really did, at one time, have back legs.[3]

Darwin himself offered a prime abracadabra on the subject:

> I can see no difficulty in a race of bears being rendered, by natural selection, more and more aquatic in their habits, with larger and larger mouths, till a creature was produced as monstrous as a whale.[4]

PANEL 16
Many Ways of Seeing Things

Darwin having had the shudders about the eye in January 1860, seems to have been happier by April. He wrote again to his friend Asa Gray: "I remember the time when the thought of the eye made me cold all over, but I have got over this stage of the complaint."

His solution in *Origin* was that "natural selection has converted the simple apparatus of an optic nerve merely coated with pigment and invested by transparent membrane, into an optical instrument as perfect as is possessed by any member of the great Articulate class." Ernst Mayr approved this, saying that there was "a correct nucleus in his claim," even thought it was "somewhat oversimplified to put it all down to photosynthesis."

George Gaylord Simpson wrote that one can observe in life on Earth today "representative stages at every gradually different level, from diffuse photosensitivity of the whole body through scattered photosensitive cells to cell plates, basins, basins and vesicles plus lenses, and so on to the fully developed image-forming eye with lens, iris, and its other complexities. These photoreceptors function splendidly at every level and do not wait to start working until the final stage is reached. They simply enlarge, refine, and to some extent change their functions as they become more complex."[20]

An illustration in Sir Gavin de Beer's *Atlas of Evolution* graphically showed the Darwinian explanation of eye evolution. But a critical reviewer aptly commented: "This mere listing of eyes from various animals, which he neglects (or is unable) to show to be related can carry no conviction for the case for evolution. It would be equally stupid to place a candle, a torch and a searchlight side by side and proceed to advance to a genealogic relationship."[21]

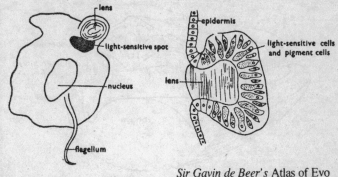

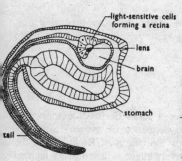

Sir Gavin de Beer's Atlas of Evolution *showed (top left) a one-celled organism with light-sensitive spot and a primitive lens, (above) a section through the eye of a jellyfish showing light-sensitive cells forming a cup-shaped retina, and (bottom left) a tadpole where the retina is formed from the lining of the brain cavity. Taken in sequence, the three diagrams give an impression of the evolution of the eye from simple to complex.*

Cautious man as he was, Darwin must have had second thoughts about this, for the sentence was quietly dropped from later editions of *Origin*. However, Sir Gavin de Beer in his *Atlas of Evolution* was not deterred. Whale ancestors, he wrote,

> had dentitions enabling them to feed on large animals, but some took to preying on fish and rapidly evolved teeth like sharks . . . Next, some whales preyed on small cuttle fish and evolved a reduced dentition. Finally the whalebone whales, having taken to feeding on enormous numbers of small shrimps, also evolved rapidly.[5]

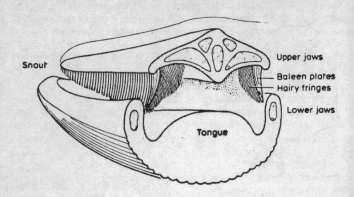

Section through whale's mouth. After the dinosaurs died out, whales evolved with astonishing rapidity.

Fanciful Scenarios

There is, as we have seen, not a scrap of fossil evidence to support this confidently stated sequence of events. Similarly unfounded guesswork surrounds the origin of flight. Thus the *Encyclopaedia Britannica* tentatively suggests that insects learned to fly because "wings arose as fixed planes extending sideways from the thorax and were used, perhaps in some large leaping insect, for gliding. Later, muscles developed, first to control inclination and then to move the wings in flapping flight."[6]

As for dinosaurs, their ability to fly is supposed to have come about because primitive swamp-dwelling reptiles, like crocodiles today, had developed strong hind legs and tails primarily for swimming. When the swamps dried up, they came on land to hunt food, where their pre-adapted legs enabled them to sprint short distances. Gradually they became two-legged, upright runners. Some remained on the ground to form the dinosaur populations. Others, "with arms freed from the burden of support and locomotion, took to gliding and developed a flap of skin stretching between arms and trunk."[7] After gliding, they learned to flap their wings, and to fly.

The supposed history of bird flight runs on much the same lines—tree-climbing led to gliding and then to proper flight. As for bats, honest bafflement is once more the order of the day. No standard explanation exists, only a frustrated amazement.

On the subject of flight, it is notable that a number of zoologists and biologists weaken their attachment to the gradualism that Darwin insisted upon. George Simpson wrote that the early stages of evolution of the bat wing "must have been many times more rapid" than after it had become fully formed[8]; which is really another way of saying that there was an inexplicably fast first stage, followed by fifty million years during which the bat wing didn't evolve at all. In the same vein, a dinosaur textbook says that the ancestors of pterosaurs "already possessed a fast-metabolizing physiology" which made the transition to flight possible, after which things remained more or less stable.[9]

Pre-Adaption

The most subtle and most recent of biological solutions to evolutionary novelties is to invoke a process called *pre-adaptation*. It is a word that seems to have two slightly different meanings. At its simplest, the case quoted in chapter two, of the bacteria that became antibiotic-resistant, serves as an example. The few mutants that turned out to be unaffected by the drug are said to be pre-adapted to the catastrophic change in their environment, and were therefore able to take over an ecological niche when their fellow bacteria were wiped out.

Reptile   **Bird**

The reptile–bird transition involved major anatomical changes, none of which have survived as fossils.

Pre-adaptation also means being lucky enough to inherit an organ that can be put to some different purpose later on. What use is a feather until it is a "proper" feather? What use is a lung that is half-developed, and cannot give you enough oxygen?

The answer comes back that if dinosaurs were warm-blooded (which is nowadays increasingly thought to be the case), then rudimentary feathers instead of scales might be very useful, for they would help to conserve heat. Half a lung, or even a quarter of a lung, might be freakily useful to a fish, for the air bubble would make it more buoyant, and it would not have to put in so much effort keeping off the bottom of the water.

In other words, incipient or intermediate organs do not necessarily work in the same way as their perfected descendants. Stephen Gould gave an answer to one of the challenges listed in panel 11 (page 62) in this way:

> The first fishes did not have jaws. How could such an intricate device, consisting of several interlocking bones, ever evolve from scratch? "From scratch" turns out to be a red herring. The bones were present in ancestors, but they were doing something else—they were supporting a gill arch located just behind the mouth. They were well designed for their respiratory role; they had been selected for this alone and "knew" nothing of any future function. In hindsight, the bones were admirably preadapted to become jaws. The intricate device was already assembled, but it was being used for breathing, not eating.[10]

This sophisticated approach toward evolutionary change has obvious uses, and offers some plausible explanations. The criticism of it that has been made is that it is too widely applied—it has become a convenient catch-all solution that is sometimes true, but also serves as a rubbish bag for all evolution's awkward odds and ends. It is hard to imagine—and certainly there is no fossil evidence for—the lucky pre-adaptations that would have twisted a bat's pelvis through 180°; or helped change the snakes' salivary glands into sacs producing some of the most virulent poisons known.

Or, to return to that puzzle of puzzles, the human eye. Here, talking about pre-adaptation is simply a way of avoiding the issue, as Stephen Gould cheerfully admitted: "We avoid the excellent question, What good is five percent of an eye? by arguing that the possessor of such an incipient structure did not use it for sight."[11]

But if not sight, what else? It is unreasonable to ask for a speculative evolutionary scenario for every single novel creature and organ that appears suddenly in the fossil record, but the most obvious and daunting ones continue to stare us in the face, unexplained.

At this point a disinterested outsider might fairly conclude that evolutionary theory has reached an impasse. In three crucial areas where neo-Darwinism can be tested, it has failed:

- The fossil record reveals a pattern of evolutionary leaps rather than gradual change.
- Genes are a powerful stabilizing mechanism whose main function is to prevent new forms evolving.
- Random step-by-step mutations at the molecular level cannot explain the organized and growing complexity of life.

Some other process, it seems, must be involved—and as we shall see shortly, there are many possibilities outside the straitjacket of neo-Darwinism.

But there is one last puzzle before coming to alternatives. "Natural selection," ever since Darwin put the words, at his publisher's request, in the subtitle of *The Origin of Species*, has become not just biology's unifying principle, but its mantra: a phrase embodying a kind of spiritual power.

Darwin himself endowed it with an almost metaphysical quality: "Natural selection is daily and hourly scrutinizing every variation, even the slightest; rejecting that which is bad, preserving and adding up all that is good; silently and insensibly working at the improvement of each organic being." Ernst Mayr compared it to a sculptor, Gavin de Beer called it a master of ceremonies, George Simpson thought it like a poet or a builder, Theodosius Dobzhansky said it was similar to "a human activity such as performing or composing music."

But is the phrase, at heart, empty? Is it anything more than a statement of the obvious?

Survivors Survive

It seems to have been the geneticist T. H. Morgan, pioneer of fruit fly research, who first spotted the problem. He wrote early in this century: "For it may be little more than a truism to state that the individuals that are best adapted to survive have a better chance of surviving than those not so well adapted to survive.''[12]

A tautology (or truism) is a self-evident, circular statement empty of meaning, such as ''Darwin was a man,'' or ''biology is studied by biologists.'' The trouble with natural selection (and survival of the fittest) is that it seems to fall into this category.

Darwin supposed that an individual creature with a particular advantage—the ''fittest among its kind'' would be naturally selected to pass on the advantage to its offspring. A horse with long legs, for instance, would be able to gallop faster than the rest, and escape from wolves or some other predator, and would survive to produce heirs. A ''fit'' creature, therefore, was one best able to carry out the functions that kept it alive—best adapted to its local environment and its way of life.

But in science, this is a speculation that can never be satisfactorily proved. The normal scientific method—the test of observation and experiment—cannot be applied to single rare events that have happened in the past. Any number of other circumstances might have led to the death of the horse with long legs. Why shouldn't the wolves have eaten it when it was young before it was able to run so fast? Or eaten old horses, so that the speed of the long-legged horse would have been unnecessary? Or might it not have died from a heart attack, brought on by exhaustion from galloping so fast?

Even Thomas Huxley admitted that Darwin had not proved that natural selection automatically produced new species, only that it ''must'' have done so; actual proof was unobtainable. He added disarmingly that even if natural selection should turn out, in the fullness of time, to be an inadequate explanation, Darwin would still be regarded by posterity as a thinker as eminent as Copernicus or Newton. In this, of course, he was right. It takes a moment or two to realize that Huxley, by dismissing in advance the need for evidence, converted a working hypothesis into an established theory without going through the accepted scientific procedure.[13]

Neither Darwin nor Huxley nor any other evolutionist of the time could define a criterion for fitness which would inevitably be naturally selected in an individual creature. Whether the long-legged horse had been fit was decided in retrospect, after it had died. Darwinism, as Darwin wrote it, could be simply but nonsensically stated: survivors survive.

Which is certainly a tautology; and tells us nothing about how species originate, as even Darwin's supporters admit. Ernst Mayr

PANEL 17
Is Natural Selection Meaningless?

Most biologists, while admitting an element of tautology in the term natural selection, nevertheless think that it still has true scientific value. Darwin discussed the problem at some length in correspondence with the American botanist Asa Gray, and wrote that "natural selection" was a convenient phrase for expressing a much more complex line of thought.

With the hindsight of our modern knowledge of genetics, the logic of Darwin's case might be expressed as follows. All giraffes alive a thousand years from now will be the descendants of giraffes alive today. Therefore giraffes alive today make up one hundred percent of the ancestry of future generations. Giraffes, however, have many different sorts of genetic makeup, and "fit" giraffes will make a numerically disproportionate contribution to the ancestry of the future population. It is not simply that they will produce more offspring, but rather that they have a better *net* likelihood of living and reproducing. So there is a statistical probability that the favored genetic types will give birth to offspring of their own kind that will survive to the age of their parents when they were born.

Stephen Jay Gould, in *Ever Since Darwin,* does not agree that Darwin himself solved what natural selection really meant—"the theory of natural selection did not triumph until the 1940s." But in its neo-Darwinian form, he thinks the term is now satisfactorily defined. His key point is:

Certain morphological, physiological, and behavioral traits should be superior *a priori* as designs for living in new environments. These traits confer fitness by an engineer's criterion of good design, not by the empirical fact of their survival and spread. It got colder before the woolly mammoth evolved its shaggy coat . . . The essence of Darwinism lies in its claim that natural selection creates the fit . . . It preserves favorable variants and builds fitness gradually.[22]

The woolly mammoth, now firmly extinct, is a curious example to demonstrate survival of the fittest.

wrote that "Darwin failed to solve the problem indicated by the title to his work. Although he demonstrated the modification of species in the time dimension, he never seriously attempted a rigorous analysis of the problem of the multiplication of species."[14] George Simpson wrote in 1964 that "the book called *The Origin of Species* is not really on that subject."[15]

Neo-Darwinist "Fitness"

But does neo-Darwinism's redefinition of the theory fare any better? The modern synthesis began in 1930 with the widely acclaimed book *The Genetical Theory of Natural Selection* by the British geneticist and mathematician Sir Ronald Fisher. Here, instead of trying to establish what makes an *individual* creature best fitted to survive in a changing environment, he looked at the population *as a whole*. The sole criterion for fitness, he wrote, is the number of offspring left by particular creatures within the population, without reference to their way of life.

Whether this has got rid of the tautology is still debated. Sir Peter Medawar and Stephen Gould think yes (panel 17). But Conrad Waddington of Edinburgh University said at the Wistar symposium:

> The theory of neo-Darwinism is a theory of the evolution of the changing of the population in respect to leaving offspring and not in respect to anything else. Nothing else is mentioned in the mathematical theory of neo-Darwinism. It is smuggled in and everybody has in the back of his mind that the animals that leave the largest number of offspring are going to be those best adapted also for eating peculiar vegetation, or something of this sort; but this is not explicit in the theory. All that is explicit in the theory is that they will leave more offspring.
>
> There, you do come to what is, in effect, a vacuous statement: Natural selection is that some things leave more offspring than others; and you ask, which leave more offspring than others; and it is those that leave more offspring; and there is nothing more to it than that.
>
> The whole guts of evolution—which is, how do you come to have horses and tigers and things—is outside the mathematical theory.[16]

We have come full circle. To put it at its mildest, one may question an evolutionary theory so beset by doubts among even those who teach it. If Darwinism is truly the great unifying principle of biology, it encompasses extraordinarily large areas of ignorance. It fails to explain some of the most basic questions of all: how lifeless chemicals came alive, what rules of grammar lie behind the genetic code, how genes shape the form of living things.

But if natural selection is found wanting, are the alternatives any better? As we shall see now, there is no shortage of people who think so.

PART TWO

Alternatives

Our faith in the doctrine of evolution depends upon our reluctance to accept the antagonistic doctrine of special creation.

—L. T. More of the University of Cincinnati,
The Dogma of Evolution

CHAPTER FIVE

Creation vs. Evolution

"The world is so full of a number of things: I'm sure we should all be as happy as kings." One of the greatest blessings of the study of God's creatures is the increasing sense of wonder and gratitude it generates. The planning and fabrication of the infinite array of beautiful animals in the heavens and on the lands and in the seas, with systems of incredible complexity and marvelous symbiosis, can only be explained in terms of an omniscient Creator. One of the greatest mysteries of human nature is the fact that intelligent scientists, familiar with these phenomena, can actually attribute them to blind chance, acting through random mutations and a random walk through natural processes operating on eternal matter. The only explanation of this strange fact is, as the Apostle says, they ". . . became vain in their imaginations. Professing themselves to be wise, they became fools . . . they did not like to retain God in their knowledge." (Romans 1:21, 22, 28.)

—Henry M. Morris, Ph.D.
Director of the Institute for Creation Research, San Diego, writing in its monthly bulletin *Acts and Facts*, November 1978

On 20 October 1972 the scientific journal *Nature*, one of the most eminent publications of its kind in the world, spoke trenchantly about "an especially foolish course" of action which the State Board of Education in California was about to undertake: to approve, in future, only those textbooks which gave equal treatment to the teaching of both evolution and the doctrine of Creation as described in the Bible.

With an air of incredulity, the writer of the leading article pointed out that "Darwinism occupies a place in science at least as strong as that of Newton's laws. No doubt there are many reinterpretations and refinements to come, but nobody in his senses can deny that the doctrine of evolution is an exceedingly powerful means of relating such a variety of phenomena that it deserves to be called the truth, if in other than scientific circles, such a term is needed.''[1]

The journal went on to offer a free subscription to *Nature* to the first ten qualified scientists, actively employed in a university science department relevant to the study of evolution, who would be prepared ''to affirm that present observations are in their opinion inconsistent with the now commonly accepted views of Earth and species evolution.''

The offer closed on 30 October (just ten days later). In the event there were just two takers: Dr. Garret Vanderkooi, Assistant Professor of the Institute for Enzyme Research at the University of Wisconsin, and Dr. Harold Van Kley, Assistant Professor in the Department of Chemistry at Saint Louis University.

This (presumably) gratifyingly small response can only have been because the closing date was premature, or because *Nature's* circulation and influence in the United States is less widespread than supposed. For there are known to be well over 700 *paid-up* members of creationist organizations there who have, at worst, a postgraduate degree in a scientific discipline; and as a correspondent to *Nature* pointed out, there are many more who prefer to keep quiet—''for them to speak out would be to invite ridicule, and probably ruin their careers. Can you blame them for keeping silent? Do you really suppose that the offer of a year's free subscription to *Nature* will tempt them to expose themselves?''[2]

The Genesis Story

In the Christian world, the most widely held alternative to Darwinian evolution has continued to be the version given in Genesis. A Gallup poll commissioned by the magazine *Christianity Today* found at the end of 1979 that of approximately 155 million Americans aged eighteen or more, no less than half ''believe God created Adam and Eve to start the human race.'' Of the thirty million people who considered themselves ''evangelicals,'' the belief was held by almost an eight-to-one margin.

What has come as a shock to the scientific community is how widespread the belief is among their fellows. In 1959, at the time of Darwin's centenary, few wise heads would have shaken in open disagreement when Sir Julian Huxley said on a television program: "Darwinism removed the whole idea of God as the creator of organisms from the sphere of rational discussion."

Yet little more than a decade later the Californian Board of Education decided in favor of modifying school textbooks to take account of the creationists' point of view, and during the 1970s the way evolution was taught in the classroom was drastically altered in many other states. Alice B. Kehoe, an archaeologist at Marquette University, Milwaukee, summed up at the annual conference of the American Anthropological Association in 1978:

> Like most Americans, I assumed that the issue of evolution versus the Biblical creation account had been demolished by Darrow's attack upon Bryan's fundamentalism at the Scopes trial in Tennessee, 1925. When, four years ago, I was asked to participate in a debate on the issue, I reluctantly agreed to what appeared to me an atavism. To my astonishment, the audience for this debate, on a November Friday night, was the largest I have ever addressed, over 1,000 persons filling a high school auditorium beyond its legal capacity. Creationism is far from dead: it is one of the most rapidly growing movements in America today.[3]

The trend is the same on both sides of the Atlantic. In Britain, membership of the Evolution Protest Movement has quadrupled in under twenty years; nearly a quarter of a million copies of its "unanswerable handbook," *Evolution: Science Falsely So-called*, have been printed. Since the war, it has attracted a sizeable number of qualified scientists to its membership, some quite eminent, such as the zoologist Douglas Dewar.

American Fundamentalism

But it is in the United States that the growth of the creationist movement has been most powerful. The rebirth of fundamentalism in respect of evolution can be dated to one year: 1963. Ten evangelists belonging to the American Scientific Affiliation (membership limited to "men and women of science who share a common fidelity to the Word of God and to Christian Faith") became dissatisfied with what they saw as a half-hearted commitment to special creation by the organization, and set up the Creation Research Society in Orange County, California.

They called themselves "scientific creationists," and they insisted that voting membership would be limited to those who had a postgraduate degree in science, and who also believed in the literal truth of the Bible. It was a timely move, for in the same year the Supreme Court ruled that it was unconstitutional to force nonbelieving children to read prayers in school.

This ruling, as members of the new society saw at once, could paradoxically be used to further their interests. One of the founders helped two women locally to petition the education authorities to "seek justice for the Christian child." Christian children, they argued, must be equally protected from having to read atheist teachings—which Darwinian evolutionary theory clearly was, since it denied the Genesis story.

The success (or as orthodox biologists would say, the nuisance value) of this first step towards creationist legitimacy can readily be seen in the textbook controversy which resulted. Meanwhile the Creation Research Society grew rapidly in strength, and moved to Michigan. In 1980 it had a thriving quarterly journal with a circulation of 2,000, its articles written by some of the 600 scientists who had qualified as voting members.

An offshoot of the society, formed after a dispute over leadership in 1970, calls itself the Creation-Science Research Center, and can be found in an apartment block in San Diego, southern California. Its director is Kelly L. Segraves, husband of one of the two women who put in the original petition. Its aim is "to reach the 63 million children in the United States with the scientific teaching of Biblical creationism," and to this end it publishes a magazine, a series of textbooks, visual aids, and gives legal advice to any parent wishing to challenge their local education board on the teaching of evolutionary theory.

In 1972 yet a further schism among the fundamentalists took place when the CSRC became divided over questions of copyright. Some of its members moved to the outskirts of San Diego and joined the Christian Heritage College as its research division. It became known as the Institute for Creation Research, and by 1980 had become far and away the most active and wealthiest creationist body in the country. Sales from its publishing associates for the previous year amounted to $354,000, and its total operating budget was $575,000. It carried out an indefatigable program of lectures, debates, newsletters, books, monitoring of science journals, media contacts, conferences, workshops, and summer institutes.

*   *   *

When I visited the Institute for Creation Research towards the end of 1978, it was not difficult to discover some of the reasons why so many people find its message appealing. It is housed in an old building in the Spanish style, formerly a monastery, set attractively beneath arid hills, and adjoining a small campus. It was rather like walking into a large family gathering united by a common heritage, with everyone enjoying a comforting secret which you are hospitably invited to share. No Bible-thumping goes on; it does not need to, for it would be impossible to find anyone out of the 200 students and twenty staff who has the least doubt that what the Bible teaches is ineffably true. The serene confidence that pervades the place is infectious. The answer to all life's problems is right there, it seems, if only you will take the simple step of joining these friendly people in their belief.

The exhibits in a small museum were in the process of some reconstruction during my visit. Shells from the beach at San Diego were compared with fossil shells, to show how little they had changed or evolved. The lack of transitional forms in the fossil records was stressed, demonstrating how God had created each kind of living thing uniquely. The marvel of a trilobite's compound eye, arriving abruptly in the fossil record during the Cambrian, was seen as evidence of God's inspired design. Fossils stretching across layers of sedimented rock showed how the Great Flood had buried them in the brief timescale described by Noah. Plaster casts of "footprints" in the mud of the Paluxy River bed in Texas, found alongside prints of dinosaurs, were said to be evidence of man and dinosaurs living at the same time. To my unpracticed eye, the footprints looked more like random indentations—but outlined in chalk, the human shape was undeniable. One print was at least fifteen inches long. Truly, there must have been giants in those times; or hoaxers in ours.

The institute's director is Henry M. Morris, who has a Ph.D. in hydraulic engineering from the University of Minnesota. Of the two dozen or so members of the advisory board or staff of the institute, almost all hold postgraduate degrees from reputable and even distinguished universities, where some continue working as professors in such disciplines as physics, biology and engineering. They are seasoned scientists, therefore, and have a good deal of contempt for other creationist bodies which accept minimal standards of academic qualification. The associate director is Duane T. Gish, who has a Ph.D. in biochemistry from Berkeley, and perhaps most nearly approximates the ideal scientific creationist as envisaged by the ICR: highly intelligent, well qualified in the areas where it is necessary to confront orthodox evolutionary theory, fluent and plausible as a speaker, unstinting with his time, and totally dedicated to the truth as revealed in Holy Scripture.

Dialogue with Dr. Gish

Considering that I believe living things have a common origin and have evolved over a long period of time, and Duane Gish doesn't, there turned out to be a surprising amount of shared ground between us. His standard lecture points out the lack of transitional fossils; regards *Archaeopteryx* as a true bird; shows how the supposed horse series exhibited in many museums is more a matter of imagination than evidence; has fun with fossil apemen; demonstrates the statistical impossibility, on any ortho-dox scientific basis, of life having emerged from inert chemicals . . . in other words, parallels many of the queries in this book. We were also at one in considering the neo-Darwinian explana-tion of major evolutionary change so inadequate that it deserves to be treated as a matter of faith.

Where we began to differ politely was over the basic strategy used by the creationist movement as a whole since 1963, which, I think, has been very subtle.

Their plan of action has been based upon the following argument. Either we have had evolution. Or we have had special creation. You have admitted that the first is unprovable, and its explanation unlikely. We, too, agree that belief in the Biblical account of Creation is a matter of faith, and that many people may think parts of it unlikely. Therefore let us put the two models side by side and see which provides the better explana-tion for the observed facts. If we regard them as two competing theories, we can even make scientific predictions and see which model verifies them and which falsifies them.

In much simplified form, I have matched the two models in panel 18. Clearly, the creationist model wins hands down. In debate after debate around college campuses in the United States, against scientists in just about every field relevant to evolution, Duane Gish and other creationists have had an overwhelming vote agreeing to the apparently simple proposition that "scientific evidence justifies the presentation of the theory of special cre-ation in public schools along with the theory of evolution as an explanation of origins."

So where does the subtlety come in? Why is the creationist argument, evidently plausible, open to suspicion?

Earth's Antiquity

First, because it is quite wrong to present creation vs. evolution as if they were the only two ways of looking at the problem—as if they were two sides of the same coin. The current *explanations* of evolution may be scientifically puzzling or unsatisfactory, but this is not to say that evolution has not occurred. The evidence from every scientific discipline that has touched on the subject shows consistently that the earth is old, is part of an even older universe, and that evolution explains why we have so many kinds of organisms and why they look so different. Radiometric dating methods confirm Earth's antiquity. Geology shows how rock formations were laid down. Palaeontology shows how there were different epochs with different life forms that ran their span and became extinct. Genetics shows how living things are related to one another, and have the potential for change.

These patient researchers do not "prove" evolution (strictly speaking, proof can be obtained only in logic and mathematics). But taken together, coming as they do from so many different viewpoints, they make an overwhelming case. Also, the way that Darwinism was found to be mistaken in various ways and replaced by neo-Darwinism, which in turn is due for demolition, says something positive about scientific method. Scientists may get stubborn about their theories, hold on to them long after their writ is run, and even conspire to present their theories as if there was nothing to be said on the other side. But history shows that in the end, as facts accumulate, a change of thinking is inevitable.

Creationist Assumptions

Not so with creationism. If you once become committed to an unalterable explanation (the Biblical one), and you are forced to fit all facts within this framework, allowing no other possibility, you have by definition become unscientific. However much sympathy we may have for belief in a Divine first cause shaping the forces that created the Universe, and however well-meaning

PANEL 18
Evolution vs. Creation: Textbook Models

Event	Creation model
Origin of life	All living things brought about by the acts of a Creator; basic plant and animal kinds complete from the beginning.
Evolution of living things	Variation and speciation limited within each kind.
Complexity of life	Sudden appearance in great variety of all living forms. Net decrease in complexity.
The fossil record	Sudden appearance of each created kind, with sharp boundaries between each kind, and no transitional forms between higher categories.
Body organs	Legs, arms, wings, etc. complete from the moment of creation.

Evolution model	Evidence
All living things brought about by naturalistic processes, originating from a single living source which itself arose from inanimate matter.	Origin of information in the genetic code not understood. Statistical "impossibility" of life arising from inert matter.
Unlimited variation, all forms genetically related.	Genetic code common to all living things.
Gradual change of simplest forms into more and more complex forms.	Second law of thermodynamics requires systems to run down and become less organized (creationist argument); fossil record shows increasing complexity (evolutionist argument).
Transitional forms link all categories, without systematic gaps.	Sudden appearance of major categories, and few significant transitional links.
Gradual evolution of organs through natural selection of features conferring added fitness.	No evidence of half-formed organs. Natural selection weeds out unfit but does not create new characteristics.

the scientific creationists may be, the straitjacket of Genesis 1-11 is so restricting that to make *all* evolutionary facts fit within it inevitably ends in a perversion of science.

Duane Gish and others of his standing are well aware of this problem, but in the end they let their faith override it. When I asked him what were the biggest difficulties for creationist science— the points in a debate which he felt least comfortable in answering—he answered after a moment's thought that it was the apparently great age of Earth as shown by the fairly recent advances in radiometric dating; and that the fossil record could be interpreted as showing ecologically complete ages—the age of invertebrates, the age of fishes, the age of reptiles, and so on up to the present. It was possible to see these as self-contained epochs succeeding one another, whereas the Bible said they were laid down during the single episode of the Flood.

Did this worry him enough ever to make him doubt his belief in Scriptural truth, I asked? "No, not at all. It just means we've got to get to work on the science of these things and find out why they do not confirm the Biblical account more readily."

In his answers to these questions in public, Duane Gish follows the normal logic for creationists: the Biblical account is correct, ergo any other finding is wrong. Thus in *American Biology Teacher* in 1973 he argued that if a Supreme Being created the world, and a catastrophe like the Flood altered it, then the evidence for radiometric dating was simply irrelevant.

Just as medieval theologians used to argue endlessly about how many angels could be placed on the head of a pin, creationists, among themselves, puzzle about the difficulties raised by geology. An ICR monthly newsletter in 1978 reported on a "significant panel discussion" held before an audience of 500 at Wheaton College, a liberal arts college outside Chicago specializing in Biblical studies. The interpretation and meaning of Genesis 1-2 was debated, particularly the question of how much time the Biblical text would allow. "The truthfulness, verbal inspiration, and the authority of the Bible were not at issue as all of the panelists, as well as Wheaton College, stated that they held to these doctrines . . . Scientific data were not involved in these discussions except as they had a bearing on Biblical interpretation."

On the one side were evangelists who felt that Genesis was to some extent allegorical, and that the six days of Creation could be regarded as eras or epochs rather than days. Duane Gish strongly contested this, according to the ICR report:

Gish said that attempts to reconcile Genesis with geology lead to numerous contradictions, even if one rejects evolution. Some of the questions Gish said must be answered are as follows. If the rock strata constitute the record of hundreds of millions of years, where is the record of the Biblical Flood? If man is several million years old, why was it that post-flood man developed agriculture and animal husbandry only a few thousand years ago when, according to Genesis, these skills were known from the very beginning? Further, why did it take so long to generate the population explosion when studies showed that a population increase of one fourth the present rate will generate our present population in just 5,000 years?[4]

Such topsy-turvy thinking (as the real majority of scientists would view it) is deeply engrained in all creationists. Here is A. J. Monty White, a chemist with a Ph.D. in gas kinetics from the University College of Wales, Aberystwyth, explaining how, having undergone a religious conversion in 1964, he finally came to see the true light:

I got married in January 1969 and it was then that I realized that the way I interpreted Genesis 1-11 was not the same as I interpreted the rest of Scripture. The reason for this was that my wife was a Creationist—she had no scientific arguments to support her belief, just the Bible—but that was enough. She would argue with me concerning my theistic evolutionary views and show me that they were not Scriptural. I realized that I was starting with evolution and trying to make Scripture fit it. Slowly I began to see that the doctrine of Creation (*not* evolution) was not confined to Genesis but occurred throughout Scripture. Indeed, I realized that much New Testament teaching was based on a literal interpretation of Genesis 1-11. Within a few months, I yielded to the arguments of Scripture—I told the Lord that although I did not have any scientific arguments to support my beliefs, I would accept Genesis 1-11 as being literal and historical. I not only believed this in my heart but confessed this with my mouth. At this time I asked the Lord to show me what was wrong with the scientific arguments of evolution and also to show me scientific evidences in support of the Genesis account of the creation and early history of the earth.[5]

The Genesis Flood

Over the course of the next two years, he says, God answered his prayers. "I sorted out dating, the fossil record, chemical evolution and much of the astronomical stuff." Two years later, in 1971, he was given a copy of *The Genesis Flood*. This major book, more than 500 pages long, written by ICR's Henry Morris and the theologian John C. Whitcomb, illustrates perfectly the strengths of the best of scientific creationist literature. It gives references for every authority cited. It has a long chapter on "problems in Biblical geology," in which the authors face up to the difficulties posed by conventional geology, such as radiometric dating, or the evidence of antiquity shown by such features as fifteen successive forests on top of one another on Specimen Ridge in Yellowstone Park. At the same time it brings out dozens of geological anomalies and puzzles ignored in most orthodox accounts: for instance, fossil graveyards, rock strata perversely out of sequence, inexplicably large formations of coal straddled by fossilized trees.

But the book's weaknesses, on any objective reading, are also readily apparent. There is the selective quotation of cautious scientific doubts—a confession of ignorance about a particular geological difficulty is cited in order to throw doubt on geological knowledge as a whole. This is a widely used technique in books and articles stemming from ICR and elsewhere. Reading creationists on the subject of ancient man, for instance, you would never gather that fossils available for study now come from a wide variety of places, and however fallible, dubious and self-seeking individual fossil finds may be (see chapter eight), they fit into a *general* pattern of man having evolved from an apelike ancestor at some point during the last six million years.

Instead, the evidence offered by creationists invariably consists of the most obvious frauds and fossil fancies, together with three or four anomalous skeletons and skulls uncovered during the latter part of the nineteenth century in strata apparently tens or hundreds of millions of years old (e.g., those at Calaveras, Castenedolo, Olmo, Abbeville, Natchez), which present individual problems for archaeologists, but can almost certainly be explained as fakes or intrusive burials.

Here, as throughout *The Genesis Flood*, we are constantly asked to accept the *least* likely solution—the opposite of both science and common sense—which demands the most parsimonious explanation for the greatest body of facts. If there is no evidence at all in the creationist's favor, guesswork takes its place. Tucked away in a footnote on page 280 is an "explanation" of why no pair of dinosaurs successfully survived Noah's custody of them aboard the Ark:

> If representative dinosaurs were taken on the Ark (presumably young ones), then it is likely that their final extinction is accounted for by the sharp changes in climate after the Flood. On the other hand, some may have persisted for a long time, possibly accounting for the universal occurrence of "dragons" in ancient mythologies.[6]

Miraculous Explanations

Elsewhere the abracadabra technique, which I complained of among neo-Darwinians, is used openly. Here the authors discuss Genesis 1:11, where God has commanded the earth to put forth grass, herbs yielding seed, and fruit trees bearing fruit of their own kind.

> One thing, however, is very significant. Plants, in order to continue to grow in the present economy, must have a soil,

water, light, chemical nutrients, etc. The account has mentioned water and light, though in a somewhat different physical context than now provided, but the soil and nutrients must also be available. As now formed, a soil requires a long period of preparation before becoming able to support plant growth. But here it must have been created essentially instantaneously, with all the necessary chemical constituents, rather than gradually developed over centuries of rock weathering, alluvial deposition, etc. Thus it had an appearance of being "old" when it was still new. *It was created with an "appearance" of age![7]*

This book, as I say, is among the best in creationist literature, and since when all else fails it can invoke the miraculous, there is really no way of arguing with its message. Some time after I had met Duane Gish in San Diego, I caught up with him again in Brighton on the south coast of England, where he was due to debate with Professor John Maynard Smith of Sussex University, a doughty neo-Darwinist.

It wasn't so much a debate as a statement of two irreconcilable points of view. The lecture theatre was packed to overflowing; Maynard Smith made a theatrical entrance, clasping his hands above his head in salutation like a professional boxer about to go fifteen championship rounds. His fans, or should I say students, roared their approval. Gish made a confident, knowledgeable speech about the fossil record and the lack of transitional links. Maynard Smith was equally trenchant, saying that as far as he was concerned there were hundreds of transitional forms, the duck-billed platypus being one of them. No vote was taken, though undoubtedly the great majority were on Maynard Smith's

The duck-billed platypus: Transitional, according to Darwinians; unique, say creationists.

side. But in England, students by and large are no longer Christian, except in name. "A tragedy," Duane Gish said sadly to me afterward.

Battle in the Courts

There can be little doubt, I think, that the surge of confidence among the creationists during the 1970s will have a lasting effect. Viewed dispassionately, safely distant from the heat of battle on the other side of the Atlantic, one should perhaps withhold judgment on whether the effects are ultimately for good or bad.

Most American scientists see them as overwhelmingly pernicious. Alice B. Kehoe finished her lecture saying she recognized that many Americans took hope and comfort from the creationists' message, and were they to keep this to themselves, it could do no harm.

But "unhappily for us non-believers, their doctrine is aggressively promulgated. Its proponents are determined to guard their children and save ours by demanding that their conceptualization of the world and man's nature be taught as a scientifically valid perspective in tax-supported institutions. Anthropologists must be alert to the maneuvers of this clever, energetic, well-organized set of Christian men who would purge the world of every belief except their own rigid Scriptural fundamentalism."[8]

In 1981, following President Reagan's cautious endorsement of their case during his election campaign, the success of the creationists in the biology textbook controversy gathered pace. Scientists who thought that Clarence Darrow's demolition of the fundamentalist case in the Scopes "monkey" trial of 1925 (see pages 181-82) had buried the teaching of creationist evolution in schools once and for all, began to find themselves on the defensive again.

Court actions in South Dakota and California took place, and more were threatened. Arkansas, which in 1968 had been the last of the southern states to repeal its laws forbidding the teaching of evolution, and then only as a result of a Supreme Court ruling in *Epperson* v. *the State of Arkansas*, put a new bill back on its books. Henceforth, state schools would be required to give balanced treatment to both Darwinian and "scientific creationist" interpretations of evolution.

The Supreme Court ruling can be circumvented, seemingly, by regarding the Biblical account of creation as science, not religion, to be treated on an equal basis with Darwinism. It is this, of course, which so angers Alice Kehoe and others who have unwittingly found themselves on the opposite side of the fence, confronted by what they see as a perversion of scientific method.

As Dorothy Nelkin of Massachusetts Institute of Technology described their objections:

> Creationism is a "gross perversion of scientific theory." Scientific theory is derived from a vast mass of data and hypotheses, consistently analysed; creation theory is "God-given and unquestioned," based on an *a priori* commitment to a six-day creation. Creationists ignore the interplay between fact and theory, eagerly searching for facts to buttress their beliefs. Creationism cannot be submitted to independent testing and has no predictive value for it is a belief system that must be accepted on faith.[9]

The Ark and Adam

Creationist teaching starts early, and while one may suppose that even the most trusting pupil must sooner or later raise a quizzical eyebrow at such comic book fantasies as the adventures of Professor Poopfossil (panel 19), it is only an extreme example of the special pleading which creationist literature aims at children. A $1.25 paperback, *The Great Dinosaur Mistake*, by Kelly L. Segraves of the Creation-Science Research Center (his son Kasey, aged thirteen, brought the 1981 court case in Sacramento, California, complaining that his teacher had wrongly taught him that he was descended from an ape), asks how Noah managed to fit two sixty-ton brontosaurus into the ark. The answer: "God probably sent two baby dinosaurs. They take up a lot less room, are easier to take care of, and have a tendency to live longer after the catastrophe."[10]

The plausibility of the ark comes in for much detailed analysis (panel 20). So does the Genesis account of Earth's first six days. Richard Niessen, who teaches fundamentalist divinity at a Christian college in El Cajon, California, wrote an *Impact* paper on the subject (presumably aimed at adults, although the same kind

THE BIG LIE - EXPOSED!

Professor Poopfossil believes in evolution. He tells people, "Once I was an amoeba so very thin. Then I was a frog with my tail tucked in. Then I was a monkey in a jungle tree. Now I am a teacher of insanity."

The Bible says that men who don't know God are "like the beasts". (Psalms 49:20) This is what people who believe in evolution are like in the spirit.

People who believe in evolution say that one animal can change into another. The Bible says they can't. Have you ever seen any of these animals? Have you ever seen a "giraffant" or an "eleraff"? How about a "pigdeer" or a "catdog"?

Of course these animals don't exist. God made it so that animals cannot change into other animals. A dog will always be a dog, never ever a cat. And, of course, a monkey can never change into a man.

When people believe in evolution they don't believe in God. They think they came from animals so they act like animals! They think God can't see them. Boy, are they going to be surprised when they die and find out God is real!

The Bible says that God created everything in only six days! He

created the light on the first day. On the second day He made the sky. The third day He made the dry land appear and gathered the waters into seas and made all the plants. The fourth day He made the Sun, Moon and stars. The fifth day He created all the sea creatures and birds. And on the sixth day He made all the beasts and man. Doesn't that sound better than the "Big Lie" of Evolution?

BM BOX 6155 LONDON, WC1V 6XX — Atten. S & R.

PANEL 20
How Did Everything Fit in the Ark?

Creationist analysis of the Biblical account of the Ark must perforce start with the precise measurement given in Genesis 6:15 "The length of the ark shall be three hundred cubits, the breadth of it fifty cubits, and the height of it thirty cubits." According to most calculations, this would form a rectangular ship of considerable stability displacing about 24,000 tons—a good size even by modern standards: comparable with a modern British light aircraft carrier such as H.M.S. *Invincible*, and nearly half the displacement tonnage of the giant U.S. aircraft carriers.

The next question is how many species had to be squashed on board. Of the one to one and a half million species living on earth today, creationists say only a relatively small proportion had to be given house room, because, of course, fish remained happily swimming about in the sea. Noah took with him only those species living "upon the face of the ground, both man, cattle, and the creeping things, and the fowl of the heaven."

A noncreationist correspondent to *Nature* said the ark should therefore contain "an aviary with 25,000 species of birds and an animal colony containing, in addition to 25,000 species of amphibians and 6,000 species of reptiles, 15,000 pairs of mammals, giving average space of less than one cubic meter for a pair of vertebrates plus food supply for one year."[13]

Kelly Segraves (see text for his suggestion that God sent baby dinosaurs to alleviate the space problem) worked it out differently, saying only 17,500 species need be accounted for. He went on to complete his calculations triumphantly:

We still have a slight problem with the figure of 17,500. If you have studied several years of biology and a couple of years of physiology you many have noticed this one fact: to preserve life upon the earth, a male and a female are required, preferably these should be of the same species. So we need to take 17,500 males and 17,500 matching females of each species, a total of 35,000 animals inside the ark. The average size of these animals is about the size of a sheep. You say, wait a minute: I have gone to the zoo and seen large animals such as elephants, giraffes, rhinoceroses, hippos, all these big animals which are larger than sheep. True, but fortunately there are also rats, mice, prairie dogs, etc., that are smaller than sheep. You can place 35,000 sheep in 146 railroad boxcars specially designed with three tiers of flooring.

We must also account for the insects. Again taking Ernst Mayr's figures, there are 850,000 species of insects. We need a male and a female of each species. Fortunately Noah did not have to figure out if he had a male and a female tse-tse fly because God brought all the animals and insects to him. We need room for 1,700,000 individual insects for the 850,000 species. Giving each of these insects 2 inches of flying space, being

very careful to pack the termites in the middle, 850,000 pairs of insects can be placed in 21 railroad boxcars. 146 railroad boxcars are needed for the animals and 21 railroad boxcars for the insects; 167 railroad boxcars in all.

The ark, figuring the cubit at the smallest known measurement, 17.5 inches, has the capacity of 522 railroad boxcars. Very simply, that means we can put all of the animals in the lower story, Noah and his family on the second story, and we have plenty of room on the third story for recreational facilities. Or you could use the third deck for animals which have gone extinct before our time, such as dinosaurs.[14]

Discussion of the problem of excreta disposal is notably absent from creationist literature.

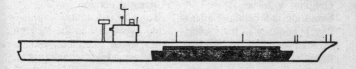

The Ark, according to biblical measurements, would be more than one third the size of the vast U.S. aircraft carrier Tirpitz, 1,092 feet long.

of imaginative thinking is involved as with the baby dinosaurs). One problem, he said, concerned the events on the sixth day: how could Adam have named all the animals in such a short period?

The answer, according to Niessen, is that Adam didn't have to name *all* the animals, only the ones mentioned in Genesis—all cattle, the fowl of the air, and beasts in the field. "This obviously eliminates most of the organisms of the earth: insects, mice, lizards and fish need not apply for the position."

Moreover, "Adam must have had an extremely high intelligence. Because Adam was capable of using 100 percent of his pre-Fall brain, he would probably have had an IQ of 1,500 or better. Furthermore, Adam did not have to learn his vocabulary: God programmed it into his brain at the moment of his creation, and he was created as a fully functioning person. It was therefore with the utmost facility that Adam named the animals that were brought before him."[11]

Evolutionary Alternatives

The flight from reason, the nuisance and expense of court cases, and the dogma returning to the classrooms may all be counted on the debit side of creationism's growing influence. On the other hand, it seems fair to say that the reappraisal of textbook evolutionary teaching has brought some gains, albeit modest ones. Here are some typical revisions accepted by the California Board of Education, all of which strike me as an improvement:

Original Version
Evolution is a central explanatory hypothesis in the biological sciences. Students who have taken a biology course without learning about evolution probably have not been adequately or honestly educated.

Modern animals that are descendants . . .

Changed version
Evolution is a central explanatory hypothesis in the biological sciences. Therefore, students need some knowledge of its assumptions and basic concepts.

Modern animals that seem to be direct descendants . . .

How do we know . . . ?	On what basis has it been concluded?
Slowly, over millions of years, the dinosaurs died out.	Slowly, the dinosaurs died out
Shortly after the flying reptiles took to the air, the early birds developed.	Birds appear in the fossil record shortly after flying reptiles.
Plants took to the land and conquered it.	Plants appeared on the land

Another gain, in the long run, may be that the creationist insistence on the scientific basis of their case will make things equally difficult for themselves *and* for neo-Darwinists. For by opening up the debate as to just what brought about the origin of life forms, they have lifted the lid off a Pandora's box brimming with alternative metaphysical and physical explanations.

If school authorities intend genuinely to present students, in a neutral manner, with the opposing claims of special creation and biological evolution, students ought to be told about creation myths from other sources, some of which are helpful to the Biblical account, and some of which are not.

. There are flood legends, for instance, in very many (though not quite all) parts of the world, which closely match Noah's story, and to some extent, therefore, support the creationist belief. But the mystery of man's origin is often dealt with in ways quite different from the Bible. Australian aboriginals believe their ancestors were animal forms that merged indistinctly with rocks and stones that have a living force of their own. American Indians, too, saw an essential unity pervading the Universe, a "Great Spirit" with the ultimate power of life and creation, but which at the same time had an accessible personality when mankind asked for help. Elsewhere people believed that man was formed from clay in a world already inhabited by animals and plants. Others teach that there is no such thing as original creation, only eternal cycles of life, death and rebirth, in which our world provides only one level of reality.

On the biological side, one comes back repeatedly to the conclusion that the creationists would not be making so much of the running were not neo-Darwinism defended to the teeth as the only viable alternative. Paradoxically, the greatest service which the creationist movement may yet perform is to spur on a basic reevaluation of the laws underlying evolution.

Colin Patterson, the palaeontologist at the British Museum of Natural History chosen to write its current book on evolution, was bombarded with creationist literature soon after the book's publication in 1978. To his surprise, he found himself impressed with some of it. "They didn't have the right answers, but they certainly asked a lot of the right questions," he has said.[12]

CHAPTER SIX

Catastrophes and Extinctions

> Posterity will admire and avenge you.
>
> —Inscription on tombstone of
> Jean Baptiste de Lamarck

Having rejected the neo-Darwinist synthesis because it is inadequate to answer these and many other questions, and rejected the creationist explanation because it cannot be argued, what, then, do we put in their place? How else do we come to have lions and jellyfish and things?

There are two basic problems which have repeatedly surfaced so far in this book, and have to be faced by any alternative theory. The first is the way that new forms of life appear very suddenly in the fossil record, usually stay there for a long while without changing much, and then die out as abruptly as they arrived. The second is whether there are universal laws of form that underlie evolution: whether lions and jellyfish evolved not just because they were good survivors, but because in a sense they were mathematically ordained to happen that way.

Although the two problems are closely linked, for the sake of clarity I am going to deal with them one after another. As it happens, they represent something of a divide within biology itself. Among critics of orthodox evolutionary theory, who throughout the 1970s grew in number and confidence, there are those who seek little more than a change of emphasis in neo-Darwinism: they accept genetic restructuring as the driving force of evolutionary change, but seek to explain how it can happen quickly rather than gradually. At the other extreme are more fundamental criticisms suggesting that evolutionary biology, precisely because of its preoccupation with genes, has traveled up a backwater—indeed, it is not yet a science at all.

Hopeful Monsters

So, genetic upheavals first. One renegade name which, in the critical literature, has noticeably been undergoing a process of rehabilitation is that of Richard Goldschmidt, a refugee from Hitler's Germany who continued his work at Berkeley. Even as the modern synthesis was being forged by such people as George Simpson during the 1930s and 40s, he saw the difficulties this posed for macro-evolution. He did not dispute conventional theory so far as gradual and continuous change within species was concerned. But he put forward a long list of evolutionary features (panel 11, page 62) that he thought inexplicable on this basis: the wing of a bird, the mammal eye, and so on.

He proposed, and thus aroused the ire of orthodox evolutionists, that these changes had happened suddenly, through "monstrous" mutations—the kind that produce fairground exhibits like two-headed sheep or stunted rabbits. He agreed that almost all of these would fail to survive; but just occasionally, a monstrosity would make the grade—and in this way a new species would emerge. He gave the newcomer a catchy name: *hopeful monster*.

He was denounced for both what he did and didn't say, but although today he rates hardly a footnote in most orthodox textbooks, researchers looking for a faster mode of evolution than the gradual accumulation of single mutations beloved by neo-Darwinists are finding in his work much to commend. "Goldschmidt's 'hopeful monster,' a mutation that, in a single genetic step, simultaneously permits the occupation of a new niche and the development of reproductive isolation, no longer seems entirely unacceptable," wrote Guy Bush, of Texas University's zoology department, in 1975.[1]

The central theme of his major book, *The Material Basis of Evolution*, is that small but significant genetic changes at the embryonic stage can give rise to large changes in adulthood.

He quotes approvingly his contemporary Otto Schindewolf, who had pointed out that those who look for missing links in the fossil record are doomed to look in vain: "The first bird hatched from a reptilian egg." Goldschmidt was one of the first to realize that the genetic system was much more complex than the one-for-one correlation implicitly accepted by most experts at the time—a gene for eye color, another gene for tail length, and so on. He embraced the notion of *rate genes*, and established that small differences in the timing of pigmentation in the embryo resulted in large differences in the color patterns of full-grown caterpillars. In general "a genetic change affecting the rate, time of inception, time of determination, range of regulatory ability of embryonic processes, may occur in a single step without requiring a rebuilding of much of the genetic material."[2]

So an entirely original monstrosity might appear on the face of the Earth in one genetic leap; and if it happened to find a congenial ecological space, could become established there as a new species. "A fish undergoing a mutation which made for a distortion of the skull carrying both eyes to one side of the body is a monster. The same mutant in a much compressed form of fish living near the bottom of the sea produced a hopeful monster, as it enabled the species to take to the life upon the sandy bottom of the ocean, as exemplified by the flounders."[3]

Embryonic Change

On the face of it, this seems obvious. The potential for change ought to be greatest when the organism is at its most vulnerable: during embryonic development, before it has become rigidly set in its adult ways.

But as elsewhere in evolutionary teaching, there is a schism about the importance of what happens in these early stages. The orthodox guardians of neo-Darwinism today come down on one side and the new wave of biologists on the other. Thus John Maynard Smith, while paying lip service to embryology in his books, has said in debate that even if and when the process of development is fully understood, it will not affect his belief that the gradual accumulation of mutations is sufficient to explain change.[4]

Stephen Gould, on the other hand, sees the mysterious processes of development as crucial to a reformulation of Darwinist theory: "The problem of reconciling evident discontinuity in macro-evolution with Darwinism is largely solved by the observation that small changes early in embryology accumulate through growth to yield profound differences among adults."[5]

Gould points out that if the high prenatal rate of brain growth is prolonged into early childhood, a monkey's brain becomes almost human in size. Similarly, if the onset of metamorphosis in a certain species of axolotl is delayed, it reproduces as a tadpole with gills and never transforms into a salamander.

"Indeed, if we do not invoke discontinuous change by small alteration in rates of development, I do not see how most evolutionary transitions can be accomplished at all. Few systems are more resistant to basic change than the strongly differentiated, highly specified, complex adults of higher animal groups. How could we ever convert a rhinoceros or a mosquito into something fundamentally different? Yet transitions between major groups must have occurred in the history of life."[6]

Chromosomal Jumps

The main objection to Goldschmidt's idea has always been that it is hard to imagine how a single monstrous mutant could have any effect on a whole population. Even if it was "fitter" than its brothers and sisters, its genetic input into the pool would soon be swamped, and the population as a whole would remain stable. We shall see later on that a number of mechanisms are being proposed to solve this problem. But this aside, Goldschmidt is increasingly being credited for seeing that large-scale transformations, rather than gradually accumulating "improvements," must have been responsible for the big evolutionary changes; and that the most likely place to find these changes is the embryo.

Within the Goldschmidt concept, U.S. geneticists have been finding other flaws with the new-Darwinist insistence that it is point mutations which count (point mutations are the single-letter changes in the genetic code which, it is said, eventually add up to instructions for new forms of life). Allan Wilson of Berkeley has identified what is perhaps the most damaging research finding: statistically, point mutations and macro-evolution seem to have nothing to do with each other.

> The rate at which point mutations accumulate seems to be the same in all groups of organisms that have been tested. That was such an astonishing result that people found it very difficult, for at least a decade, to believe. Point mutations accumulate in this steady way, whereas anatomical change occurs at very different rates in different groups of organisms.
>
> For instance, frogs evolve very slowly, but their point mutations change at the standard rate. Or take chimpanzees

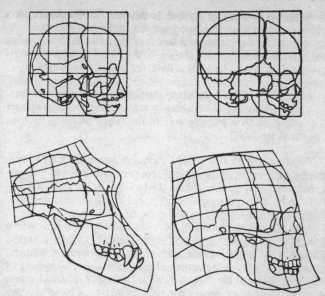

The skulls of chimpanzees (left) and humans are very alike at the embryonic stage (above). But growth into adulthood (below) takes them into dissimilar forms. The rules underlying the geometric change are not understood.

and people. At the point mutation level you could scarcely tell us apart. Something else must have made us evolve so differently.[7]

Rather than mutations as traditionally considered, it may be alterations in the chromosomes that lead to alterations of life forms (chromosomes are the pairs of rodlike structures that appear within the cell when it divides: they contain the genes). This is sometimes called the theory of chromosomal speciation, and is being put forward by others besides Allan Wilson. Guy Bush, for example, has pointed out that chromosomes evolve much faster (by a degree of magnitude) in most mammals than in most other vertebrates.[8] Michael White, of the Australian National University of Canberra, whose book *Modes of Speciation* is the definitive recent academic text on the subject, comments that under a microscope you can readily see chromosomal differences between closely related species of higher mammals. At gene level, chimps and ourselves may be difficult to tell apart; but our chromosomes are obviously unalike. Population geneticists such as Mayr, he thinks, have not examined in depth

> the extent to which structural rearrangements of chromosomes, such as inversions and translocations, have played a special and perhaps a primary role in the origin of many or even most species.[9]

Together, the idea of hopeful monsters formed in the embryo, through a form of chromosomal reorganization not yet properly understood, may give evolution the kick up the backside it needs to fit in with the palaeontological jumps the fossil record shows. But is even this formula radical enough?

A satisfactory evolutionary theory has also to show how whole groups of mutations happened simultaneously. When a reptile learned to fly, it involved the harmonious development of feathers, growth of wings, strengthening of certain muscles, lightening of bones, upheaval of the digestive system, and other related changes. A small number of geneticists believe that to achieve this we have to accelerate change still further, and resurrect another name which, like Goldschmidt's, is anathema to all true Darwinists: Jean Baptiste de Monet de Lamarck.

History dealt unkindly with Lamarck, both in his lifetime and afterwards. Born in 1744, he became acquainted as a young man with the philosopher Jean Jacques Rousseau, who inspired him to move into biology. In middle age he was a professor at the Museum of Natural History in Paris, but it was not until he was sixty-five that he produced the book which can now be seen as the first substantive challenge to the creationist view of evolution: *Philosophie Zoologique*, published in 1809, just half a century before Darwin's *Origin of Species*.

Unlike *Origin*, Lamarck's thesis (which he expanded and amended in a textbook on invertebrate zoology six years later) never found acceptance in the scientific community. He died, blind and in great poverty, at the age of eighty-five. His passing was marked by a scurrilous "eulogy" from his former colleague, the great French anatomist Baron Cuvier. Even when published in an expurgated form by the French Academy of Sciences, with the most offensive remarks deleted, it had the effect of discrediting his work further.

Since then he has been increasingly cast in the role of demon king to Darwin's benevolent dictator. "The discarded and disreputable theory of Lamarck" is a quite mild neo-Darwinist dismissal of his work.[10] He has been ill-served even by his supporters, who have insisted on only one aspect of Lamarck's complex treatise: the inheritance of acquired characteristics. The notorious Lysenko, who dominated Russian biology during the Stalinist years, equated this supposedly Lamarckian principle with Marxism in an attempt to increase crop yields and breed better livestock. He failed amid appalling scenes of suppressed scientific evidence and the elimination of his opponents.

Thus Lamarckism, in both popular and scientific circles, has simply come to mean that parents can pass on to their offspring attributes picked up during their lifetime. A blacksmith's children will grow up with stronger arms, a giraffe's with longer necks; crossbred wheat that survived in Siberia would grow steadily more cold-resistant in succeeding generations.

Lamarck, however, made a clear distinction between the kinds of traits that might be passed on. If it resulted from the direct action of the environment (heat, light, injury, mutilation) it was of no importance to evolution. He would, I am sure, have agreed with the many experimental findings which support this, but which are held against him. If you chop off the tails of mice for a hundred generations you still will not breed a species of tailless mice.

But he believed firmly that use and disuse of an organ, *in pursuance of a creature's basic needs*, could in the end have an effect on evolution. He cited the life of a mole:

> Eyes in the head are characteristic of a great number of different animals, and essentially constitute a part of the plan or organization of the vertebrates . . . Yet the mole, whose habits require a very small use of sight, has only minute and hardly visible eyes, because it uses its organ so little . . . Light does not penetrate everywhere; consequently animals which habitually live in places where it does not penetrate have no opportunity of exercising their organs of sight . . . It becomes clear that the shrinkage and even disappearance of the organ in question are the results of a permanent disuse of that organ.[11]

This is the truly contentious part of Lamarckism. Does a living thing somehow have within itself, consciously or unconsciously, a way of responding actively to the environment and shaping the future of its species accordingly? If so, it would go a long way toward solving the two basic conundrums of evolution I outlined earlier—its speed, and the need for a number of mutations to take place simultaneously. Can we find, within an organism, a response system that accelerates genetic change in this way?

Penetrating Weismann's Barrier

Until very recently, the answer would have been a resounding no. Such a suggestion runs counter to one of neo-Darwinism's most hallowed principles: Weismann's doctrine.

To remind ourselves (see also page 54), this says that there is a complete and total "barrier" between the cells involved in reproduction (germ cells) and those in the rest of our body (somatic cells). That is to say, our bodies may respond and adapt and become injured during our lifetime without affecting the genes which will eventually be passed on to our children. We can acquire leathery, weather-burned skins in an equatorial desert, or grow layers of fat to keep out the cold in the frozen north; but our germ cells are impervious, unchanged except by occasional random mutations within them. The barrier was proposed by the German biologist August Weismann shortly after Darwin's death, and on the whole it has stood the test of time well. Weismann's barrier, one critic has written, is "by far the most important tacit assumption in contemporary biological thought"[11]; and another, "it is my view that the central block to any consideration of the inheritance of acquired characteristics can be directly traced to Weismann's doctrine."[12]

Small wonder, then, that when at the end of the 1970s a number of experiments began to cast doubt on its validity, the bedrock of evolutionary theory began to tremble.

The attack on Weismann's barrier came from an unexpected source: the Ontario Cancer Institute in Toronto. Here three young scientists, in 1978, decided to apply their professional and specialist knowledge of immunology and biochemistry to Darwinism's genetic impasse. Ted Steele, then twenty-nine, is Australian; Reg Groczynski, two years older, is British; so is their friend Jeffrey Pollard, from a different group in the institute.

"There was a feeling we all had that traditional evolutionary theory just didn't smell right," Steele has explained. "We were all agreed that the neo-Darwinian explanation as it stood didn't solve such problems as multiple linked mutations. We also thought it was making some basic assumptions that needed to be questioned, including its refusal to consider a form of Lamarckian inheritance. In immunology, developmental genetics are a good bit more flexible and complicated than most population geneticists imagine. We thought it might provide us with a way of breaking through Weismann's barrier.''[13]

The upshot, little more than a year later, was a cautious paper by Steele and Groczynski in the academically renowned *Proceedings of the National Academy of Sciences* in the U.S., and a short book, written by Steele and edited by Pollard, entitled *Somatic Selection and Adaptive Evolution: On the Inheritance of Acquired Characters*. In it he inquired provocatively of Weismann's doctrine: "Can it be replaced by something more useful and rational? I am aware of no direct observations which unequivocally prove that genetic information (DNA, RNA) *cannot* pass, in principle, from differentiated body cells to ova or sperm."

Indeed, the Toronto experiments showed the reverse (panel 21). Adopting a well-known technique of immunology, young mice were "persuaded" to tolerate foreign tissue antigens from another kind of mice. (Normally, as we know from transplant surgery, tissue from another body is rejected.) *This tolerance was inherited by the offspring.*

"Clearly, we were seeing inheritance of an acquired characteristic," according to Steele. "The cells of one strain of mice learned to accept the cells of another strain of mice, and this learnt tolerance was passed into the germline.

"The technique is relatively simple. If you inject a large quantity of foreign cells—about ten million—into a mouse's bloodstream soon after it is born, when its immune system is not complete, it does not reject them, as it would later in life. What seems to happen is that the genes in the body cells of the mouse are altered so as not to respond to foreign cells, and throughout the rest of the mouse's life they regenerate somatically in this slightly different form.

"What we didn't realize was that this change could be inherited. According to Weismann's doctrine, it shouldn't—it should stay in the body cells and there would be an end to the matter. But when you examine the mouse's children, and grand-children, you find the tolerance to foreign cells passed on in just the numbers conventional Mendelian theory would expect—that is, roughly half the children, one quarter the grand-children.''[14]

PANEL 21
Can Mice Inherit "Tolerance"?

In a classic series of experiments in the 1950s, Sir Peter Medawar established that an animal's immune system develops very early in life. The purpose of the system is to recognize foreign material—for instance, viruses and bacteria —and then fight it and eliminate it. The immune system does not destroy any of its own body tissues, which is what is meant by "tolerance."

Sir Peter Medawar found that you can alter the immune system somewhat by giving an animal massive doses of foreign material soon after birth. If a pure-bred strain of mice (say, strain A; see illustration) is injected with body cells from another strain (strain B), it accepts these, and later in life cannot distinguish between A and B. It is "tolerant" to B cells.

A normal adult mouse would reject the cells—it would be "intolerant."

In the Gorczynski/Steele experiments, the tolerance to B cells was passed on to two generations of offspring, suggesting that the tolerance had somehow penetrated Weismann's barrier. Attempts to repeat the experiment by other researchers showed ambiguous results. Steele thinks the work supports his theory; the researchers don't.[30]

Inheriting acquired characteristics in this way does not seem to be a precise process. "It is more like a general strategy," Steele has said. "It is a way of helping a species to respond rapidly and flexibly in times of environmental crisis."[31]

Lamarckism Reborn

The potential importance of the work that the group pioneered may be judged from some of Ted Steele's reviews. The *New Scientist* called it "the first concrete evidence that evolution does not occur only by natural selection." Sir Karl Popper found it the year's most exciting book. Sir Peter Medawar, the Nobel prizewinning biologist in London to whom Steele went to have

Female [Strain A] Male

[Strain B]
Source of foreign tissue

Strain A babies injected with strain B tissue

Male   Some adult strain A's do not reject grafts of strain B cells

Normal strain A female   Male

B-cell tolerant males mated with normal strain A females

Adult offspring grafted with strain B cells

Predicted by Darwinian theory - B cells rejected

Observed - B cells accepted

The Toronto experiments in inheritance of an acquired characteristic—in this case tolerance to foreign cells.

the results confirmed, said that if this happened satisfactorily, it would represent "one of the landmarks in the history of biology . . . it will establish a principle of the utmost importance for evolution."[15]

In the event, Ted Steele performed another successful experiment and had the results published in *Nature*; results from independent researchers were less conclusive, perhaps because they did not follow Steele's precise technique.[16] In 1981 he returned to his home country in search of funds to continue the experiments (fittingly, he comes from the town of Darwin), certain in his own mind that a breakthrough had been achieved.

The daunting title and technical prose of the Steele/Groczynski paper ("Inheritance of Acquired Immunological Tolerance to Foreign Histocompatibility Antigens in Mice") give little indication of the private enthusiasm Steele shows for its implications. In particular, he is excited at how the accepted rate of evolutionary change can be increased phenomenally.

"We have observed genes changing direction in a single generation. That is astounding. If multiple changes in the body can register in the germline in such a short time, you can stop thinking about evolutionary developments taking millions of years, and talk about a few generations instead. In controlled conditions, we have seen the rate of molecular change in mice speed up 10,000 times above normal. With coupling effects, it's feasible the speed-up could be one hundred million times."

Steele does not visualize this new form of Lamarckism as a precise process. "The kind of picture I have in mind is, say, a drastic environmental change, or an epidemic disease. To meet this challenge, the immune system which exists in all living things starts working overtime. Genetic variability is increased dramatically, and other unforeseen changes, not just immunological ones, start happening simultaneously and are passed back into the germline.

"When we did our experiments with mice, it wasn't just immunological tolerance which was passed on to the next generation. There were unaccountable changes of fur colour as well—genetic changes we had no reason to expect, but which were obviously associated.

"Perhaps we shouldn't have been so surprised. The immune system is part of the blood system, and also has a memory. Imagine, a human regenerates a pint of blood a day! That's hundreds of millions of new cells being turned over daily, many of them, in conditions of stress, with substantial genetic alterations. And they go everywhere in the body.

"It is known for sure in molecular biology that certain kinds of viruses can display the potential to transfer information from one body cell to another. All we are suggesting is that more information can be transferred than we suspected, and that some of it gets back into the germ cells as well.

"If we are right, it means that an organism both adapts to its own environment and anticipates, to some extent, the future environment of its offspring. This is an instinctive process—but it's truly Lamarckian, if you go back to what he actually wrote, rather than what he is said to have written.''

It is also a much more creative and flexible approach to evolution than neo-Darwinism allows, for it accurately portrays a dynamic interplay between organisms and their environment at all levels, including the molecular.

Maternal Effects

Moreover, if we return once more to the emerging embryo, the findings provide a possible mechanism for all manner of maternal effects that have been observed in biology—that is, the way that the behavior of the mother affects the embryo. Smoking or excessive alcohol drinking by a human mother causes harm, for instance. The nature of a mother's food, any illness she contracts, even altered behavior under stress—all these are known to have an effect, and the memory of them will be imprinted on genes in the embryo's body cells.

How many of these effects then become heritable (with the embryo growing up and passing them on in turn to its offspring) is still debated. But some are, which means that new information is being genetically assimilated by a mechanism that almost certainly preempts natural selection. Similarly, there is a whole range of acquired characteristics known as *dauermodifications* that have been observed in single-celled organisms, insects, plants, and even a few mammals. Here an environmental stimulus causes a change, and the change persists for several generations even when the environment returns to normal.

Indeed, using only modest speculation, Steele's work suggests a solution to two classic Darwinian puzzles: how did ostriches get their calloused knees, and we humans the thickened soles on our feet?

An ostrich has callouses on its legs where it kneels on the ground. To an extent, these develop during its lifetime, in much the same way as a sailor's palms grow tough if he continually wrestles with ropes. Similarly, if we walk barefoot for long enough, our soles become hard and leathery.

But in both cases, *the thickening begins before birth*—inside an ostrich's egg and inside a human womb. We are born with the job already half done. The thickening process is dictated by our genes.

So how could this have come about? Lamarck would have had no difficulty in replying that it was an obviously useful attribute acquired gradually over many generations, and passed on to our children. But so long as Weismann's barrier was sacrosanct, such an explanation was not allowed. Now that the barrier has apparently been penetrated, the easy answer may yet turn out to be the right one, even if the detail is not immediately understood. (The neo-Darwinist explanation, panel 22, is tortuous in the extreme.)

It would be idle to pretend that a rehabilitation of Lamarck, even if overdue, provides all the answers to evolutionary puzzles. Steele himself sees it as an invaluable extension of neo-Darwinist selection principles rather than a total replacement. He thinks the main contributions are to provide for a very much faster mutation rate, and a way for evolution quickly to track environmental changes.

However, the picture he builds up challenges another of the tacit assumptions which has accompanied evolutionary theory: that on the whole, the history of Earth has been as stately as evolution itself. Worldwide climatic changes are seen as gradual and insidious, keeping step with the gentle pace of mutation.

Steele seems to suggest that evolutionary jumps haven't usually happened like that. His neo-Lamarckian mechanisms come into play when the environment cuts up rough—when the stimulus it gives to an organism is "intense and chronic," as he puts it.

By implication, he brings to life another of Darwinism's specters—another word banned from the canon of respectable investigation: *catastrophism*.

Have the upheavals in Earth's past—floods, plagues, earthquakes, and bombardments from the cosmos—been more important than the grindingly slow processes of erosion and continental drift?

Catastrophism and Uniformitarianism

Catastrophism became discredited in the nineteenth century at about the same time as creationism, and for much the same reasons. In the debate between Church and Science, it found itself on the wrong side.

Catastrophism, in the mind of the general public at the time, meant simply a belief in Noah's Flood, the single great event which had changed the face of the Earth. To scientific creationists of the time—Baron Cuvier, Louis Agassiz and others—catastrophism was rather different. Examining the tortured geology of the rocks, with its folds and upheavals, and evidence of past extinctions, they concluded that catastrophes had struck Earth repeatedly.

None of these scientists seriously believed in the Flood, nor that any of the cataclysms had been supernaturally (i.e., Divinely) caused. They also found absurd the recent date for the Creation in the margin of most Bibles of the time (and still annotated in many Gideon Bibles found at hotel bedsides today): 4004 B.C. Instead, they saw in the fossil record an Earth of great antiquity, and one which had in various epochs been very different from the modern world.

It was their misfortune, and subsequently geology's too, that the two forms of catastrophism—one founded on religious faith but the other a remarkably accurate reading of what we now believe to have taken place—became inextricably confused. Against them was put a supposedly more scientific doctrine, uniformitarianism, which won the day and has kept its deadening hand on the subject more or less ever since.

Uniformitarianism had its birth with the publication of Charles Lyell's *Principles of Geology* in 1830. A lawyer, he had studied for a time under the creationist geologist William Buckland, subsequently Dean of Westminster, who taught that "Geology is the efficient auxiliary and handmaid of religion," and who saw widespread evidence of "direct intervention by a divine creator." Lyell sought an alternative that avoided the miraculous, and said that geological history could perfectly well be understood by observing processes still going on—the drip of rain, the wind-blown grains of sand, the pulverizing descent of glaciers. "The present is the key to the past" became the geologist's catchphrase, and in due course this innocuous statement of method duly took its place in post-creationist science.

But Lyell went on to infer something much more arguable: that the processes on Earth had *always* been much the same as now, and that rates of geological change had been uniform through time. It was a concept eagerly adopted by Darwin, who

How Did the Ostrich Get Its Horny Knees?

"The ostrich may develop callosities first as a non-genetic adaptation. But the habit of kneeling, reinforced by these callosities, also sets up new selective pressures for the preservation of random genetic variation that may also code for these features."[32] Stephen Gould's explanation is the orthodox neo-Darwinian one, known usually as the *Baldwin effect*, or (the term coined by Conrad Waddington) *genetic assimilation.*

What is deemed to happen is a three-stage affair. 1, a novel adaptive response (in this case, calloused knees) arises during embryonic development or childhood because of a new environmental stimulus (say, drought, bringing rough and stony ground on which the ostrich kneels). 2, as the environmental stimulus persists (prolonged drought over several generations), the response becomes "canalized"—i.e., it deepens in intensity and becomes more or less uniform throughout the population. 3, after some generations, the response becomes assimilated, so that it arises even in the absence of the stimulus.

John Maynard Smith described the process succinctly:

The combined effects of adaptation during development to environmental stimuli, canalization of development and genetic assimilation are to mimic Lamarckian inheritance without involving any process not known to occur.[33]

This is so close to what Lamarck himself suggested as to be virtually indistinguishable. Mae-Wen Ho and Peter Saunders complained:

To say that genetic assimilation merely mimics Lamarckian inheritance rather than it represents a Lamarckian phenomenon, is but to engage in a fine point in semantics. Similarly, to argue that a process cannot be said to be Lamarckian unless it violates the central dogma is to interpret Lamarckism in a sense which bears little resemblance to the original theory . . . Maynard Smith appears to be saying that neo-Darwinism explains evolution in terms of the natural selection of random mutations *plus* all other processes which are not known *not* to occur. We could hardly argue with such a formulation. But it is not surprising that many opponents of neo-Darwinism consider the theory to be nonfalsifiable and hence unscientific.[34]

needed slow and steady environmental changes for natural selection to work in finely graduated steps; and with the triumph of Darwin's book, this aspect of uniformitarianism became accepted too.

Ever since, it has dominated the study of geology to what can be seen as a pernicious degree. "Catastrophism became a joke, and no geologist would dare postulate anything that might be termed a 'catastrophe' for fear of being laughed at," according to Derek Ager, Professor of Geology at the University of Swansea.[17] Almost any issue of standard academic journals bears out what he says. Here is Neville George, Ager's predecessor at Swansea, in *Science Progress*:

> It might be said that evolution as we know it is a lucky accident that reflects, over 500 to 3,000 million years, a general uniformity of environmental stimulus rarely becoming so violent, at least on a worldwide scale, as to be intolerable to most major kinds of animals and plants. Only during glacial periods, recurrent a mere half-dozen times in Earth's history, and never completely worldwide in glacial onset, have environmental conditions been wholly disastrous over large parts of the Earth's surface to both animals and plants.[18]

Velikovsky's Theory

The geological establishment's antipathy to catastrophism was reinforced during the 1950s by the work of the maverick multidisciplinary scholar Immanuel Velikovsky. This Russian émigré, drawing on evidence from ancient texts around the world, rewrote astronomical and Biblical history with an audacious theory: Venus, he said, was a fairly recent addition to the planetary system, having been ejected from Jupiter and then circled Earth on numerous occasions before settling in its present orbit. Two close approaches, he said, were accurately described in the Old Testament and elsewhere in the accounts of floods, debris from space, a day when the sun did not set, and so on.

Vividly, he quoted Mayan documents which, in his interpretation, referred to "a cosmic catastrophe during which the ocean fell on the continent, and a terrible hurricane swept the earth. The hurricane broke up and carried away all towns and all forests. Exploding volcanoes, tides sweeping over mountains, and impetuous winds threatened to annihilate mankind, and actually did annihilate many species of animals. The face of the earth changed, mountains collapsed, other mountains grew and rose over the onrushing contact of water driven from oceanic spaces, numberless rivers lost their beds, and a wild tornado moved through the debris descending from the sky.''[19]

Velikovsky died in 1979, still forcefully defending his thesis against the massed ranks of scientists who thought it mad. A few academics were prepared to concede that in certain respects (e.g., his redating of Biblical history) there was something to be said for him, and the respected *Journal of Physics* printed an intriguing paper (panel 23) which suggested that Earth had repeatedly spun upside down during its past, causing exactly the kind of apocalyptic effects that Velikovsky had described so dramatically.

The pros and cons of the debate aside, there is no dispute that Velikovsky catalogued an enormous number of geological events which were certainly catastrophes of one sort or another: fossil graveyards (e.g., of mammoths in Siberia), giant boulders swept hundreds of miles from their source, mountains thrust spectacularly out of shape, cities destroyed. The origin of Earth's huge coal deposits (panel 24) seems to owe at least as much to catastrophes as to sedimentation.

But were all these events, as he seemed to imply, the result of Venus swooping dangerously close to Earth? Geologically, the evidence is said not to be good—''I have really tried to come to terms with the Velikovskyans, but without success,'' Derek Ager says. ''They have so far failed to produce anything that fits in with my own observations in geology. I positively begged them to produce some geological evidence but at that point the correspondence ceased.''[20]

So Velikovsky in his lifetime had two contrary achievements to his name: (a) he established in the minds of his readers and followers that catastrophes were endemic in Earth's history, and (b) through his imaginative, outrageous, and highly popular theory about the cause of them, he ensured that the subject was banned from respectable scientific discussion. It was curiously like a rerun of the arguments a hundred years before, when Baron Cuvier (whom Velikovsky cited approvingly) painstakingly catalogued Earth's catastrophic past, which promptly (because he was identified with the creationists) removed catastrophes from mainstream geological thinking.

PANEL 23
Can Earth Turn Upside Down?

An abiding scientific mystery is why Earth's magnetic field periodically reverses itself, so that the north pole becomes the south pole, and vice versa. Rock samples taken from deep sea drillings have a magnetic "fingerprint" on them showing in which direction the magnetic field lay at a given time; and they have demonstrated that field reversal takes place seemingly at random intervals, on average 10,000 to 100,000 years apart.

This is not just of academic interest, for the field reversals are associated with profound changes in life on Earth. They have been linked to the onset of ice ages, and to waves of extinctions. One function of the magnetic field is to keep at bay the gamma rays from outer space that are lethal to living things. It has been speculated that the extinctions are the result of the field declining to nil before establishing itself in the other direction. During the years when it is weak or nonexistent, a bombardment of cosmic radiation might decimate life forms.

But this theory suffers considerable problems, not the least being the extinctions have often included many underwa-

ter creatures, who are shielded by the sea from radiation regardless of the magnetic screen. Also, recent research suggests that the reversals may take place over a much briefer time-span than was previously thought. There was a reversal, known as the Mungo event, about 30,000 to 40,000 years ago, that lasted for no more than 2,000 years. Another (the Gothenburg event) happened about 13,000 years ago and was even briefer. Analysis of Greek and Etruscan vases suggests there may even have been a reversal as recently as the eighth century B.C.

The most radical theory to explain these mysterious incidents was published in the prestigious *Journal of Physics* in 1978.[35] Peter Warlow, a British physicist and mathematician, proposed that it is not the magnetic field which reverses itself, but the Earth turning upside down within it. He backed up this extraordinarily bold proposal with mathematics that were equally challenging. These are based on his observation that because Earth is not perfectly cylindrical (it is fatter at the equator), it is potentially very unstable as it spins. A passing asteroid of sufficient size and nearness would exert enough gravitational pull to make Earth topple quickly over. The entire process could take place in as little as a day:

> To turn the Earth upside down, with all the attendant havoc such an action can produce, I have shown that a mere three hundredth of the Earth's rotational energy will suffice—and that is only borrowed for half a day. In the following half day most of the energy is put back where it came from! This staggering reduction in energy is achieved very simply ... To divert that little piece of energy from one rotation to produce another, we only need to set a planet sailing erratically around the Solar System.[36]

As an example of a similar movement taking place, he cites the plastic "tippe-top" to be found in some boxes of cookies. Here, because it is unequally weighted inside, it refuses to stay upright when spun; it wobbles and turns upside down. Friction from the table has the same effect as gravitational pull on a spinning Earth.

Suggestions that the Earth's axis has changed, or tilted, have been made by other writers, notably Charles Hapgood[37], and the idea is central to Velikovsky's work. Critics have always been able to point to the apparent physical impossibility of this happening—too much force, or energy, seemed to be required. Now that Peter Warlow has overcome this problem, the ensuing catastrophic events on Earth appear more plausible: floods, dramatic climate changes, typhoons to destroy forests, hurricanes of sub-zero temperatures cold enough to freeze a Siberian mammoth in its tracks while chewing a buttercup[38], tidal waves carrying hundreds of thousands of drowned creatures with them and piling them up in fossil graveyards.

Because the world, according to Warlow's calculations, would go on spinning at the same speed, a human observer lucky enough to be away from the worst of these effects (in the middle of a large land-mass towards the equator, for instance) would not feel the earth shaking beneath his feet. The most startling effect would be to observe a double-length day (or night) after which the Sun would rise from the west. Evidence of this was collected by Velikovsky from many sources, and Warlow believes that Plato in the classical world, and the Hopi Indians in America, both accurately described a tippe-top world and its effects.

The consequences of the world tipping upside down, he

says, "satisfy the geological, astronomical, historic and legendary data. The actual equatorial axis for the secondary rotation can be established by mathematical analysis of the Earth's shape *or* from ancient stories of the pattern of the behaviour of the Sun and stars *or* from analysis of the pattern of sea-level changes *or* from the distribution of Flood legends *or* from the position of the extent of the last ice age. All of these totally separate routes yield one single answer."[39]

Neo-Catastrophism

During the 1970s, a rehabilitation of the subject began to gather pace. In part this was due to space exploration, which reminded us that the cosmos was anything but an empty void. It sounded increasingly improbable that Earth could have avoided contact with the occasional massive interplanetary body like those that had blasted craters on the surface of the Moon and Mars. Scientists speculated whether vast near-circular geological formations such as the Gulf of St. Lawrence, 380 kilometers in diameter, might not be fossil craters of giant meteorites; they were given the name "astroblemes." An extraterrestrial origin was at last agreed for tektites, mysterious pieces of glass, commonly pear-shaped, found in great numbers in Australia and elsewhere. Sir Fred Hoyle, the British astronomer, produced statistical evidence suggesting that many virus diseases had their origin in space.[21]

This picture of an Earth periodically bombarded, irradiated, and infected is consistent with some of the major gaps in the fossil record, which Professor Norman D. Newell, of the American Museum of Natural History in New York, has called "crises in the history of life." Looking at the geological ladder (page 5), he has identified them as having happened at the beginning and end of the Cambrian, and the ends of the Devonian, Permian, Triassic, and Cretaceous, plus a number of other times somewhat less well marked.

"The stream of life on earth has been continuous since it originated some three or four billion years ago," he wrote in 1963. "Yet the fossil record of past life is not a simple chronology of uniformly evolving organisms. The record is prevailingly one of erratic, often abrupt changes of environment, varying rates of evolution, extermination and repopulation . . . Mass extinction, rapid migration and consequent disruption of biological equilibrium on both a local and a worldwide scale have accompanied continual environmental changes."

For tens or even hundreds of millions of years, certain groups of animals and plants predominated. "Then, after ages of orderly evolution and biological success, many of the groups suddenly died out. The cause of these mass extinctions is still very much in doubt and constitutes a major problem of evolutionary history."[22]

Derek Ager agrees with this interpretation. "Changes in the past have often been both episodic and sudden," he said in a presidential address to the Geologists' Association in 1976. His slim book *The Nature of the Stratigraphical Record* is an entertaining demolition of uniformitarians who look upon today's erosion and sedimentation rates as a trustworthy guide to ancient geological history. "We do not yet know enough of the present to use it fully as the key to the past."

Punctuated Equilibria

Stephen Gould (Professor of Geology at Harvard) and Niles Eldredge (like Newell, at the American Museum of Natural History) are two other eminent teachers of evolutionary theory who can loosely be described as neo-catastrophists and who have attained positions of influence. Together they have proposed, since 1972, that gradualism should be replaced by what they term *punctuated equilibria*. "The history of evolution is not one of stately unfolding, but a story of homeostatic equilibria, disturbed only 'rarely' (i.e. rather often in the fullness of time) by rapid and episodic events of speciation."[23]

Eldredge came to this point of view almost reluctantly, in the course of researching a doctoral thesis on the evolution of trilobites. The pattern that emerged confounded his expectations of gradual change. "It wasn't like that. Over a period of several millions of years there was almost no anatomical change at all, except an increase in size. There were small, subtle alterations to the eyes, but even these came in a herky-jerky way, not gradually as classical Darwinism would predict. And this is what seems to happen right through the fossil record—long periods of stasis and then the sudden appearance of new forms."[24]

PANEL 24
How Rapidly Was Coal Formed?

Theories about the origin of coal aptly illustrate the uniformitarian vs. catastrophist debate. Coal is composed of compressed, carbonized vegetable matter, mostly formed in the Carboniferous period dated to 370 to 280 million years ago. It exists in enormous quantities—15.3×10^{12} tonnes, according to one textbook[40] —and can be found at depths of up to 1,800 meters.

Few except creationists (who see all coal as having been formed during the Flood) doubt that the gradual accumulation of sediment in swamps is a natural explanation for coal. Peat can be seen forming today; temperature and pressure increase with depth, and thus it seems to fit the theory that peat was transformed by these means over millions of years into coal.

But the explanation may be only superficially plausible. It is an uncomfortable fact that nowhere in the world today is there a known peat bed where coal is being formed. So for the theory to stand up, you have to postulate that during the Carboniferous period, climatic conditions were different, and more favorable.

Also there is a problem in explaining the vast quantities of vegetation required to produce the huge coal deposits. As Dr. Heribert Nilsson, Professor Emeritus of Botany of Lund University in Sweden, wrote:

> From the point of view of the amount of material available, the results must be considered highly improbable. A forest of full-grown beeches gives material only for a seam of 2 cm. It is not unusual that they are 10 metres thick, and such a seam would require 500 full-grown beech forests. Whence this immense material? How was it deposited all at once? Why did these masses of organic material escape decay, why was it not completely decomposed?[41]

Although Nilsson's conclusion was no doubt colored by his firm creationist beliefs, there is considerable evidence from other sources that coal can be, and was, formed rapidly rather than gradually:

1. During the construction of a railway bridge in Germany in 1882, reports were made on the condition of wooden piles rammed into the ground and compressed by overhead

blocks. Their centers had been transformed into a black coal-like substance chemically the same as anthracite. "From all available evidence it would appear that coal may form in a very short time, geologically speaking, if conditions are favorable," according to the respected coal authority, E. S. Moore.[42]

2. Whole trees—even forests of trees—are often found buried upright and intact within coal seams. Derek Ager wrote:

We do from time to time find evidence, in all parts of the stratigraphic column, of very rapid and very spasmodic deposition in the most harmless of sediments. In the late Carboniferous Coal Measures of Lancashire, a fossil tree has been found, 38 feet high and still standing in its living position. Sedimentation must therefore have been fast enough to bury the tree and solidify before the tree had time to rot. Similarly, at Gilboa in New York State, within the deposits of the Devonian Catskill delta, a flashflood (itself an example of a modern catastrophic event) uncovered a whole forest of *in situ* Devonian trees up to 40 feet high.[43]

There are also many indications that coal was formed through violent incursions of the sea. "Coal balls" of matted plant and animal remains, ranging from the size of a fist to massive examples weighing a ton, sometimes include marine fauna such as sponges, molluscs and corals. A tiny marine worm, *Spirorbis*, usually less than one-eighth inch long, never found in freshwater deposits (e.g., bogs) today, is widespread in coal seams, mixed in with plant debris.

Velikovsky's vivid scenario, though unlikely to be the whole truth, therefore seems to be more accurate than it is given credit for: "Apparently the coal was not formed in the ways described. Forests burned, a hurricane uprooted them, and a tidal wave or succession of tidal waves coming from the sea fell upon the charred and splintered trees and swept them into great heaps, tossed by the billows, and covered them with marine sand, pebbles and shells, and weeds and fishes; another tide deposited on top of the sand more carbonized logs, threw them into heaps, and again covered them with marine sediment. The heated ground metamorphosed the charred wood into coal . . ."[44]

In terms of years, it is near impossible to tell just how short and sudden were these catastrophic episodes, in which one life form was wiped out and another took its place. In geological time, they could have been an instant, and still been longer than the 50,000 years or so that *Homo sapiens sapiens* has been around. Gradual but snowballing effects of a disrupted ecological food chain might have taken that length of time to decimate life; on the other hand a meteorite impact that tipped the world upside down would have globally lethal effects within days, hours, or even minutes. Locally, a drought can change an environment out of recognition in a period of years. All are catastrophes for the life forms there. The varied suggestions for the extinction of the dinosaurs' 150-million-year rule on Earth are typical of the range of possibilities: chronic constipation at one end, and clouds of cosmic dust at the other (panel 25).

Survival of the Luckiest

Perhaps the most far-reaching effect of this revival of catastrophist thinking has been the dawning realization that mass extinction makes a nonsense of natural selection as a "creative" force. David Raup of Chicago's Field Museum has calculated that in the half-dozen major extinctions, *up to ninety-six percent of all life forms were destroyed.*[25]

Now if only four percent of living things managed to survive such cataclysms, the question of fitness is largely irrelevant. It is much more a question of chance. Instead of survival of the fittest, you get the survival of the luckiest.

The worst catastrophe, everyone agrees, was at the end of the Permian period some 225 million years ago. It established the ancestry of most of today's life forms, for since then there has been little change in the basic pattern of types. But as Stephen Gould has observed, it wasn't necessarily the best-adapted Permian plants and creatures that lived through the disaster:

> If anywhere near 96 percent of species died, leaving as few as two thousand forms to propagate all of later life, then some groups probably died and others survived for no particular reason at all. There are few defences against a catastrophe of such magnitude, and survivors may simply be among the lucky four per cent . . . our current panoply of major designs may not represent a set of best adaptations, but fortunate survivors.[26]

PANEL 25
What Destroyed the Dinosaurs?

Some sixty-five million years ago, there was an extinction of massive and dramatic proportions. The event shows up so clearly in the geological record that it has been taken to mark the end of the Cretaceous (the age of dinosaurs) and the beginning of the Tertiary (the age of mammals). About seventy percent of all living species died out. No land animal heavier than twenty-five kilograms is thought to have survived the catastrophe; nearly half the species of flowering plants disappeared; so did half the marine organisms ranging in size from tiny floating single-celled plants to strange shellfish and squidlike creatures.

The puzzle, so far as a uniformitarian explanation goes, is that there is no physical difference in the rocks on either side of the event. Whatever happened, happened quickly, and seemingly left no trace. The explanations put forward in various discussions of the subject were little more than speculations, all easily refuted.

1. "The dinosaurs over-evolved—they developed bodies too massive and cumbersome for their environment, and died of a kind of racial senescence."[45] But this was certainly not true of all dinosaurs, some of which were small and even furry; and it fails to explain the suddenness of the event.

2. "Mammals, in competition with dinosaurs, eventually defeated them."[46] Several versions of this have been put forward, following the standard Darwinian approach. While it is true that mammals ultimately replaced dinosaurs, there is no evidence that this happened as a result of unsuccessful competition—rather, the reverse. While dinosaurs ruled the Earth, mammals were extremely successful. Suggestions such as mammals upsetting the breeding patterns of dinosaurs by eating their eggs, which have been seriously advanced by Darwinian biologists, show the lengths to which invention must reach if a gradual and uniform solution is sought.

3. "An ecological change put the dinosaurs at a disadvantage, and they quickly succumbed to starvation or disease."[47] Here again, a number of fanciful explanations have been put forward that owe more to imagination than to observed fact. Perhaps poisonous mushrooms appeared

at the same time as flowering plants, and destroyed the dinosaurs by parasitism; perhaps the flowering plants themselves developed potent alkaloids to deter creatures trying to eat them, causing digestive upsets among the dinosaurs; perhaps flowering plants increased the amount of oxygen in the air, and the dinosaurs were unable to cope with it; perhaps the dinosaurs were wiped out by an unknown disease. But no account of this sort explains satisfactorily all the associated extinctions.

Once, however, the possibility of a catastrophe is admitted, the solutions become more plausible. For a while, the theory of Dale Russell, of the National Museum of Natural Sciences in Ottawa, came close to being accepted.[48] He suggested that a supernova, in which a star larger than our Sun suddenly runs out of fuel and begins to collapse, emitting colossal amounts of lethal radiation, had occurred. Even 100 million light years away, it would have disrupted the protective upper atmosphere, exposed the Earth to cosmic rays, and brought about a marked and prolonged drop in temperature. Larger plants and creatures would be unable to cope with the cold; smaller ones, and less highly evolved marine creatures, could survive and adapt.

Latterly, the favored explanation is even more catastrophist in nature. It is proposed that a giant meteorite, a few thousand tons in weight, struck Earth with the force equivalent to 100,000 million tons of TNT. The blast from the crater, mixed with meteorite dust, would create a dense, choking cloud around Earth, screening it from the heat of the Sun, and causing a chilling drop in temperature.

The physicist John Gribbin wrote:

> Volcanoes today do much the same thing on a smaller scale; the dust veil from a major meteorite impact could have blocked off enough sunlight to stop photosynthesis in many plants, killing them and destroying the food on which animals depended, while at the same time triggering a brief ice age, harsh climatic conditions to reduce still further the number of survivors.
>
> Small animals, needing less food, and the creatures of the shallow seas, keeping warm in the water, ought to survive such a disaster best—and these are, indeed, exactly the species whose remains are found after the Cretaceous-Tertiary boundary event, showing that their ancestors survived the disaster.[49]

Gribbin has also described the likely point of the meteorite's impact—not on land, where the scar would certainly still be visible, but in the ocean:

> To a large meteorite hurtling in from space, the ocean waters are scarcely any more hindrance than the atmosphere, and the energy released in the impact is sufficient to punch a hole through the crust of the sea floor, which is much thinner than continental crust.
>
> This would expose the molten rock below, and ocean water rushing into the red-hot crater would be turned to steam, spreading a blanket of cloud, as well as meteoritic dust, around the globe.[50]

Somewhere on the mid-Atlantic ridge, a vast geological fault line where the continents of America, Europe and Africa once met, now parted by continental drift, would be an ideal place for the meteorite to have landed and caused the effects it did, and for all traces to have vanished. Just one may remain: Iceland, seated astride the rift, and formed of exactly the volcanic material which would be expected. Scientific support for the theory comes from measurements of traces of iridium in rock formations marking the sixty-five million year old event. Iridium, a rare heavy metal, is far more abundant in meteorites and asteroids than in the Earth's crust, and there are exceptional deposits at the time of the death of the dinosaurs. Moreover, the nearer to the north Atlantic the measurements of iridium have been made, the greater they are.

The theory, first proposed by Luis Alvarez of the University of California at the annual meeting of the AAAS in January 1980, is much the most plausible and comprehensive to have been put forward so far. What has not yet permeated into scientific discussion is that it is in striking accord with the "tippe-top world" theory described in panel 23. Deep sea cores of sedimentary rock brought up by the U.S. National Science Foundation's drilling ship *Glomar Challenger* showed, according to the expedition's leader, "the best record ever collected of the reversal of the Earth's magnetic field that coincided with the extinctions between the Cretaceous and Tertiary periods."[51]

Mathematically, a meteorite impact of the size suggested would be enough to turn the world upside down, according to Peter Warlow's calculations. In which case the catastrophe now being contemplated by geologists would have even more devastating effects.

So far in the list of alternatives to neo-Darwinism we have had hopeful embryonic monsters; new species by way of chromosomal change; acquired characteristics passed on from parents to offspring; and catastrophes which cause mass extinction leaving a small number of lucky survivors.

If you combine all these ideas, and boil the result down to its essence, a new scenario for the origin of species begins to emerge. In as few words as possible,

A severe environmental crisis accelerates embryonic restructuring, and isolated mutants survive.

The emergence of the first land animals accordingly might be written:

In the wake of a disaster (probably global) a large number of amphibious creatures were thrown far up on shore and became stranded. Many died of starvation or injury; but many also survived, and among the mothers, the multiple effects of a changed diet, stress, prolonged exposure to a new climate, and acquired immunity to virus diseases led to intense genetic pressure on their unborn young. Chromosomal changes led to hopeful monsters in their thousands emerging from new-laid eggs. The vast majority were stillborn, or impotent, or failed to make an impact because the chromosomal change was ''bred out.'' But just occasionally, through harem-type breeding in isolated populations, the chromosome change was perpetuated. After a few generations, several varieties of ''monsters'' became viable— new species which then proliferated over a largely unpopulated globe.

This account, which is implicit in most of the alternative theories which were being put forward during the 1970s, is profoundly un-Darwinian in almost every respect.

It supposes, first of all, that speciation happens in a single dramatic step, and quite possibly a lone individual is involved.

Next, it says that *relaxation* of competition is what enables the new species to emerge and survive. The first land animals enjoyed, presumably, an empty land surface and an absence of predators. There was room for experiment in ecological space.

This is the opposite of the way neo-Darwinists think, for they assume that in crowded populations, when competition for survival among the fittest is intense, natural selection also operates intensively and new forms evolve. The first reptiles did not need to be ''fit'' in this sense at all, for competition among each other must have been virtually nonexistent.

Next, it supposes far more interaction between the creatures and their environment than the austerities of chance and necessity involve (chance mutations, the necessity for them to give the creature an advantage). Instead of the random mutation of "hopeful" genes selected impassively by an unfeeling environment, we see creatures, to repeat Ted Steele's words, adapting to their environment and anticipating the environment of their offspring.

Finally, and crucially, this account of the birth of reptiles downgrades the role of genes as the primary source of evolutionary change—and here is where I am going to draw my arbitrary line through the ranks of the theorists: genophiles on the left, genophobes on the right.

Patterns of Evolution

Most of the dissenting geologists and palaeontologists we have met in this chapter would, I think, accept that Darwinism in one form or another works at the level of molecular biology, and that this "explains" evolution (Stephen Gould may be an exception, but then he is a dissenting biologist too).

Their invaluable contribution has been to look at the rocks and fossils with unblinkered eyes and establish a *pattern* of evolution—sudden jumps often accompanied by massive catastrophic extinctions. We may even predict that, sooner rather than later, this pattern will become the new orthodoxy, so reasonably does the Ager/Gould/Eldredge school present its view. Derek Ager told a critic of his presidential address that he was not disputing theories, simply reporting what existed: "I was only saying that the following are the facts we find in the rocks if we do not allow hypothesis to interfere with observation. We do *not* see lots and lots of species all gradually changing from bed to bed; what we do see is a species persisting through a certain thickness of strata and then suddenly being replaced by something else."[27] And on another occasion:

> Nothing is worldwide, but everything is episodic. In other words, the history of any one part of the earth, like the life of a soldier, consists of long periods of boredom and short periods of terror.[28]

Geology provides the time dimension within which biologists must work to provide the *mechanics* of evolution. Historically, it was inevitable that attention should first be paid to the mystery of the genes. Genes can be counted, and analysed, and tampered with. There is no doubt that different species have different gene structures, and that changing a gene can change the life form. Science finds it hard to operate in a theoretical vacuum, and genetic theory has given scientists data that can be measured and tested.

But for all the scientific advances that have taken place as a result of genetic research, the theory has been tested to destruction so far as its effect on evolution is concerned. It is hard to disagree with what Professor Steven M. Stanley of Johns Hopkins University, Baltimore, told the U.S. National Academy of Sciences in 1975:

> The reductionist view that evolution can ultimately be understood in terms of genetics and molecular biology is clearly in error. We must turn not to population genetic studies of established species, but to studies of speciation and extinction in order to decipher the higher-level process that governs the general course of evolution.[29]

It is this "higher process" that seems to have been lost sight of during three quarters of a century's obsession with Mendelian genes. The mathematics of a genetic change are an aspect of evolution, but surely not the whole of it. Mathematicians, physicists, chemists, and even biologists who take a wider view, have concluded that behind life itself lie marvelous and complex processes that are slowly revealing their secrets; and it is these on which evolution ultimately depends.

CHAPTER SEVEN

Patterns of Life

> Some contemporary biologists, as soon as they observe a mutation, talk about evolution. They are implicitly supporting the following syllogism: mutations are the only evolutionary variations, all living beings undergo mutations, therefore all living things evolve. This logical scheme, is, however, unacceptable: first, because its major premise is neither obvious nor general; second, because its conclusion does not agree with the facts. No matter how numerous they may be, mutations do not produce any kind of evolution.
>
> —Pierre Paul Grassé,
> *Evolution of Living Organisms*

One persistent criticism directed against population geneticists is that their exclusive concentration on genes, molecules and cells as the basic biological unit has made them lose sight of a larger wonder: nature's creations as a whole. Peter Saunders is a mathematician from Canada who teaches at Queen Elizabeth College in the University of London. He and his wife Mae-Wan Ho, a biologist who helped draw up the course on evolution at Britain's Open University, have published a number of scientific papers, notably in the *Journal of Theoretical Biology*, expressing discontent with the contemporary approach.

"The thing about neo-Darwinism is that it provides a very simple explanation which is probably wrong and certainly insufficient," Saunders has said. "Almost the most surprising thing is that anyone takes it seriously at all. Heredity, variation, natural selection—of course they happen. Genetic changes also. But geneticists then go on to say that these things are a necessary and sufficient explanation for all evolutionary change—that's all there is to study.

"We find this very hard to take. It says nothing about the origin of variations. It says nothing about whether some variations are inherently more likely than others. All changes are held to be possible and all equally likely, so they don't look at the animal at all, they just examine the mathematics of genetic change. They don't look at the way an organism is built, they don't look at the way an organism develops—in fact, they neglect virtually all of biology.''[1]

The principal objection seems to be that neo-Darwinism is strictly a theory of *genes*, whereas the phenomena it seeks to explain are *forms*: insects and crabs and whales and rhododendron bushes. The theory assumes you can deduce the organism from the genes. But nobody has ever done so, and the more complex and contrary the genetic mechanism is shown to be, the more it becomes evident that it can't be done.

Laws of Form

So is there another way, in science, of approaching the problem? It turns out that there is. The answer is simply to turn the problem on its head. Instead of concentrating on the strings of atoms that make up individual genes, you look for the unifying process that enabled the atoms to put themselves together in this pattern in the first place. Instead of looking at minute variations between one gene complex and another, or one creature and another, you search for the features that are common to all— similarities, not differences. You try to perceive the basic shapes, patterns and possibilities—an intangible system perhaps even more mysterious than the genetic code, but unquestionably more fundamental.

Like catastrophism, this *morphological* approach (morphology is the study of form) had its origins in pre-Darwinist biology, and continued to attract the minds of a number of quizzical scholars during the subsequent years of Darwinist supremacy. Indeed, Baron Cuvier himself was among the early morphologists who sought to identify regularities in the diversity of adult and embryonic forms.

The German biologist Hans Driesch, who helped keep their work alive at the turn of the nineteenth century, summed up their aim: "to construct what was typical in the varieties of form into a system which should be not merely historically determined but which should be intelligible from a higher and more rational standpoint."[2]

Darwin's followers, by contrast, concluded that new forms were produced entirely by chance. Rigorously simple versions of neo-Darwinism say this even today: all gene mutations are equally likely, therefore the number of forms is also theoretically limitless.

In other words, in the creation of forms, anything at all could happen, and just did: lions, jellyfish, etc. are nothing more than a bundle of atomic parts and random events. Hans Driesch commented wryly: "Thus Darwinism explained how by throwing stones one could build houses of a typical style."

Organizing Principles

Yet at every level, it is obvious that there is something more to it than that. Life itself must obey some organizing or self-organizing principles, for otherwise there would be nothing to distinguish it from nonlife. Why should cells invariably divide into two (with rare exceptions that prove the rule)? Why not three, or seventeen, or any other number that comes up in the Darwinian lottery?

There is also the phenomenon, troublesome to neo-Darwinism but highly relevant to this deeper discussion, of what are known as *homologous organs*. These are similar structures which span whole classes of creatures and have persisted in the same basic form despite millions of generations of mutations. Some similarities can be distinguished only by the professional anatomist: the

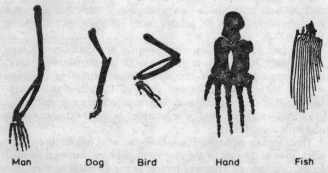

Man   Dog   Bird   Hand   Fish

Homologous organs. There seems to be an underlying law of form which makes the same pattern perform so many different functions.

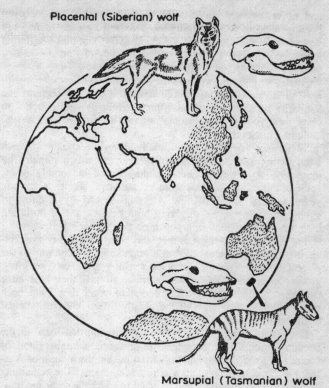

Placental (Siberian) wolf

Marsupial (Tasmanian) wolf

Skeletons of Canadian and Tasmanian wolves are almost identical, despite the two species having evolved separately on opposite sides of the planet.

jaw common to all reptiles, the inner ear of mammals. Others are more easily recognizable: segmentation in insects, or the tetrapod limb in vertebrates.

This latter is a classic textbook example of nature persuading one structure to do several jobs. Why should the leg of a horse, the wing of a bird, the arm of a man, and the flipper of a whale all be built the same way when serving quite different purposes? If the fittest adaptation were chosen by a gradual accumulation of mutations, you would have expected an organ used for flying and an organ used for running to have finished up—or even begun—looking totally dissimilar. It is not even as if they are governed in each case by the same gene complex—different genes, seemingly obeying an underlying law of form, somehow produce the same basic structure (panel 26).

Again, why should we have the phenomenon of *parallel evolution*? The most striking example of this can be seen in the similarity between marsupials (those mammals with pouches to carry their young) and placental animals (most other mammals, including ourselves).

Their joint ancestors were small, shrewlike creatures which lived during the Cretaceous, when the Earth was dominated by dinosaurs. As the continental plates drifted inexorably apart, some of these little animals were left stranded, about 150 million years ago, on what is now Australasia; and by a quirk of evolution, their manner of giving birth became one of nurturing their newborn progeny in a pouch on the mother's belly.

The extraordinary thing is that, this oddity apart, the marsupials of Australia have evolved in a remarkably similar way to the placental mammals in the rest of the world. Wolves, cats, squirrels, ground hogs, anteaters, moles and mice all have their look-alike counterparts, in spite of the millions of years they have been raised apart, and in spite of the greatly differing environmental challenges that each has had to meet.

Darwinists shrug off the problem by saying that some of the uncanny resemblance between, for instance, a timber wolf of Canada and the Tasmanian wolf from the southern coast of Australia is because they share a common ancestor. The remaining similarity is because they had to adapt to much the same environments and occupy much the same ecological niches—"Tasmanian and true wolves are both running predators, preying on other animals of about the same size and habits."[3]

It is the familiar catchall explanation that actually explains nothing. Why should the Tasmanian wolf not have, for instance, more teeth, or bigger eyes, or longer claws, or larger ears, or any of a myriad other evolutionary possibilities? In all likelihood the climate in Australia has been, at times, markedly different from the rest of the world during the millions of years involved; so too would have been other environmental factors—food supplies, epidemic diseases, number of predators, and so on. It seems highly unlikely that natural selection of chance mutations, if this were the supreme governing principle, could ever have arrived at the same near-identical solution, given the differing circumstances over such a huge timespan. Simply, it is far more probable that there is some kind of wolf "blueprint" that determines they shall evolve as wolves. As Arthur Koestler has put it, there are certain built-in rules that permit a great amount of variation, but only in limited directions on a limited number of themes.[4]

Embryonic Development

In embryology, there is emphatic support for this. Conrad Waddington, the distinguished biologist whom we met in an earlier chapter as chairman of the rancorous Wistar symposium, when engineers and computer scientists challenged the geneticists, spent most of his working life in this field. Although he always, at least in public, proclaimed himself a Darwinist, his contribution to the evolutionary debate can clearly be seen as following the tradition of the dissenting morphologists. He was sure that there were only limited opportunities for an evolving creature:

> The development of an animal under the influence of its genes is obviously an intricate and carefully controlled process, which normally leads to a rather standard and invariant end result. A mutation of one or even several of the genes concerned can either disrupt the process completely, so that the animal dies, or it can produce effects only of a limited character.
>
> If, for instance, you have a set of machine tools for producing a reciprocating internal combustion engine, changes in these tools can alter the character of the engine—the diameters of the cylinders, the stroke of the pistons, the clearances of the valves, and so on; but they can scarcely at a blow start producing a turbine or a jet. Similarly, with a developing biological organism, mutation is anything but omnipotent to produce changes in any conceivable direction.[5]

Waddington used a vivid metaphor to illustrate how, he believed, a growing organism was channeled in a certain way. He pictured hills and valleys that made up an *epigenetic landscape* (epigenesis means the origin of an entirely new structure during embryonic development—e.g., an embryonic lizard growing a wing). He pictured the embryo flowing relentlessly down a gentle slope that

PANEL 26
How Did the Giraffe Grow Its Long Neck?

The evolution of the giraffe, the tallest living animal, is often taken as classic evidence that Darwin was right and Lamarck wrong. The giraffe evolved its long neck, it is said, because natural selection chose those animals best able to feed off the highest treetops, where there is most food and least competition. A Lamarckian explanation—that the habit of reaching upwards for food gradually stretched the neck and elongated the legs over a number of generations—is deemed unnecessary and biologically impossible.

What is seldom pointed out, however, is that the giraffe is an exceptionally good example of a "biological blueprint" which directs that the form of mammals is restricted within certain parameters. A full-grown male giraffe may grow to a height of six meters or more, over half of this being the length of its neck. But it has no more neck bones than most other mammals, including man —it is just that the seven cervical vertebrae have been greatly elongated. The same is true of the legs.

Whether the lengthening happened gradually or rapidly cannot be judged from the scanty fossil record. A few fossilized leg bones and vertebrae have been found in related mammals during the Miocene (twenty to fourteen million years ago) which, according to Dr. Roger Hamilton of the palaeontology department of the British Museum of Natural History, show a tendency to long legs, and to longer necks than the deer which were common at the time. But they are not directly ancestral to modern giraffes, which appear plentifully in the fossil record about two million years ago, since when they have not changed much in shape or size. So what little evidence there is is consistent with Gould-Eldredge's punctuated equilibria (page 136); it is also possible to believe, as Roger Hamilton does, that in sites such as the Baringo area (a lake site in the Rift Valley in Kenya), "it should be possible to find intermediate forms. I'd be surprised if they weren't there. It's just lack of work."

Looking at the lifestyle of giraffes, it is hard to see that the traditional Darwinian selection pressures of competition for survival in conditions of overcrowding, and predation by other species, have very much to do with their extraordinary shape. Female giraffes are up to a meter less in height than the males (and young giraffe, of course, less still). So they feed off different heights of tree, and there is plentiful foliage left at a lower level. We cannot know that conditions were the same in the past, but the "need to survive by reaching ever higher for food" is, like so many Darwinian explanations of its kind, little more than a *post hoc* speculation.

Similarly, the observable fact is that the giraffe has virtually no enemies. The lion is the only wild animal that will attack a giraffe, and then only when made desperate by hunger. The lion, in any case, often comes off second best; a giraffe's hoof, which can strike in any direction, can kill a lion with a single blow.

Overall, the story of the giraffe's evolution is entirely consistent with the ideas being mooted in this chapter: that there are underlying laws of form and development that give much variation within certain restrictions; that individual species can express these laws in a somewhat predetermined way (the giraffe has a "tendency to long neck and legs"); that evolution may develop rapidly (the "sudden" appearance of the modern giraffe); and that these basic evolutionary drives have a momentum of their own, largely unaffected by considerations of competition.

---

led to adulthood and ultimately old age and death. This was its general, predetermined genetic pathway.

> However, we have also to take account of the fact that different parts of the embryo develop into different organs—liver, kidney, brains, muscles and so forth. This situation can be described by supposing that superimposed on the general slope is a radiating system of valleys, which direct some trajectories to move along toward the kidney, another set to move along toward liver, and so on.[6]

Waddington pictured the emerging organism, in normal circumstances, taking the "easy" pathways—that is, along the bottom of the valleys. This, he said, was how a species remained basically stable, the organs developing in the same way and the resulting animals looking much alike.

However, there was a sort of game going on between the

organism and the environment. The organism was forever exploring new possibilities as its environment changed—going up the hillsides until the biological equivalent of gravity drew it back down again. And occasionally, the environmental pressure might be so intense or prolonged that the organism was pushed clear over the top of a hill and down the other side into a new developmental pathway. The adult organism, if it survived, would emerge in a different form.

You can see this happening in nature. If, for instance, you give embryo fruit flies a sharp heat shock, you can "kick" them into a new developmental route. They emerge from the egg in a different form—fatter, or with different shaped wings, or any of a number of mutant possibilities including greater longevity. If it happens to a single individual, the change is not passed on to the offspring: it just lasts one generation. But if enough fruit flies undergo the *same* transformation, there is a way in which the effect can become "canalized," overcome the usual genetic stabilizing effect, and ultimately establish itself within an entire population.

Two Conundrums

The importance of this is manifold. It shows that Waddington's idea of hills and valleys and developmental pathways is correct, and provides the beginnings of an answer to the two basic conundrums I identified in the previous chapter: the speed of evolution, and the apparent need for a number of mutations to happen simultaneously.

For since all individuals in a population have much the same pattern of embryonic development, they will react to a perturbation in much the same way. You will have not just Goldschmidt's hopeful monster, but a whole generation of monsters arriving in one step. By the same token, if alternative forms are lying in wait within the genetic system, just biding their time until an environmental disturbance calls them into action, you do not need large numbers of simultaneous mutations, but just a single crucial one.

Waddington was never very specific about the mechanisms involved in turning his metaphor into reality. He died in 1975 before the neo-Lamarckian revival had gathered momentum. Believing profoundly, as he did, that it was the interplay between a developing organism and its environment which shaped

the end result, there is no doubt that Ted Steele's experiments
would have excited him.

Mathematical Catastrophes

But everything Waddington wrote also implied belief in a deeper
mathematical process which guided the development of life forms,
and here, during the latter years of his life, came growing
support from a number of theoreticians in other disciplines.
There was, for instance, the "catastrophe theory" of the brilliant
French mathematician René Thom.[7]

This says basically that there are certain forms in nature,
whether in biology or in physics, which are *preferred*. They are
not the only ones which can occur, but they are more easily
produced, and therefore more commonly found, than the others.

The reason for this lies in some deep and extraordinarily
complex theorems. These concern certain sorts of differential
equations which go to the heart of physics and chemistry. Scientists who have delved into them are convinced that they have the
potential to describe the fundamental physical and chemical
processes which create the "preferred" forms; they could even
turn out to be perfectly simple, except that there are so many of
them. The early stages of the growth of an amphibian embryo, and
the reflection of light from the surface of a cup of coffee, are in
most respects very different, but the same pattern can be seen in
both. Thom's work says why.

Moreover, according to Thom there are not just preferred
forms, but also preferred changes of form. If we look at one of
the small number of forms that Thom says are most common, we
may not know how it is produced, nor whether it will eventually
evolve into something else. But if it does evolve, Thom's mathematics show what it is likely to change into.

Both aspects of Thom's work give mathematical support to
Waddington's picture. It also downgrades the role of the genes in
a way which we shall find happening repeatedly from now on.
Instead of mutations determining the variation of form, the change
is, in a way, automatic: the new form was implicit in the old
one.

Laws of Creation

Mae-Wan Ho and Peter Saunders find this mathematical approach exciting. "Biology is riddled with theories that are *given*—there's a point at which you are supposed to stop asking questions and accept that you can go no further.

"With the creationists, you are given the premise that God made the world, and that's that—don't inquire further. Morphologists like Driesch began by asking the right questions, but when they got into difficulties, they invoked a vital life force, or some vague equivalent, which was supposed to be beyond investigation. With Darwinists it's the genes that are irreducible—don't ask how they got their structure and their information in the first place, they were just given it.

"Now, with the work of Thom and others, you can at least begin to see that there are basic physico-chemical rules which govern life and the evolution of life, just like everything else in the universe."[8]

Brian Goodwin of Sussex University, who once worked under Waddington and has become one of the world's foremost mathematical biologists, agrees: "There are laws of creation. There must be. They are the framework within which ordered phenomena can make their appearance.

"But neo-Darwinism doesn't tell you this. It claims to be the unifying theory of biology, but it's nothing of the kind. It's a theory of stability—the capacity of forms to persist through genetic inheritance. But it tells you nothing about the generation of form.

"If someone came to you and said he had a theory about aircraft stability and how to test it, but he couldn't tell you how aircraft were made, you would reply that this was interesting, and even useful, but very limited. It doesn't allow you to understand the principles whereby aircraft are constructed.

"Even in engineering, where stability is very important, you can't say it's supremely important. It tells you how much stress a bridge will stand, but it doesn't actually tell you how to build a bridge.

"The language of the new biology, when it is properly formulated, will be rigorous, scientifically and mathematically, but it will have a totally different flavor from contemporary biology."[9]

The Evolutionary Vision

René Thom himself has said little about biological evolution, although he is widely quoted by scientists searching for the laws of creation. In recent times, the most radical assault has come from a handful of theoreticians who subscribe to what they term the *evolutionary vision*.

Before launching into an outline of this revolutionary line of thinking (it challenges Darwinism at least as profoundly as Darwin himself challenged the Church), I should offer a word of warning. The evolutionary vision stems largely from the work of Ilya Prigogine and his school. He is Professor of Chemistry and Theoretical Physics at the Free University of Brussels, and was a Nobel prizewinner in 1977. Like Thom's, his work is expressed in mathematical equations so difficult as to stretch even the best of minds, and quite beyond lay comprehension. So in giving this nonmathematical account, I shall inevitably oversimplify, and fall short of communicating the sense of pure wonder and satisfaction that his conclusions give to all his colleagues.

Moreover (for these are early days yet) the evolutionary vision is speculative. It does not have the authority of, say, quantum mechanics, itself the subject of endless dispute. It may not even, at least at this stage, be testable in the accustomed scientific way. I asked a hard-line physicist friend, Eduardo Balanovski, who has a Ph.D. in solid-state physics from Imperial College in London, whether he thought Ilya Prigogine's seminal papers in *Physics Today* were valid in the sense defined by the science philosopher Sir Karl Popper—that is, of putting forward bold predictions which were capable of being falsified by future experiment.

Balanovski, who has followed Prigogine's work for a number of years, thought not: (a) the papers were based on a set of assumptions that were by no means proved, and (b) so many variables had to be built into the mathematics that he didn't see any way they *could* be proved. "In many ways, Prigogine is as much a mystic as a physicist or chemist. He has a view of life which may well be right, but does not altogether lend itself to our existing scientific methods—in other words, to be blunt, it is not science as we know it."[10]

That said, however, it is difficult to overstate the potential of the evolutionary vision, and the excitement it has created. A symposium at the 1980 annual meeting of the American Association for the Advancement of Science, held in Los Angeles, was entirely devoted to the subject, and attracted overflow audiences from a wide variety of scientific disciplines. Speakers tackled with zest neo-Darwinism's most intractable puzzles, and offered theoretical solutions to many of them: how life emerged from inanimate chemicals; why life forms become increasingly more complex; why evolution seems to take place in quantumlike leaps.

Above all, the audience relished the attempts to unite physics and biology—the great divide which Prigogine has sought to bridge. The enthusiasm was so infectious that a euphoric remark was overheard afterward: "This is the closest that science has yet come to describing why life exists."

The fundamental riddle which Prigogine and his colleagues have been attacking is one which has plagued science since the last century: how and why does life, in a universe of ever-increasing disorder, evolve in the opposite direction? What are the rules of life that enable eggs and buds to transform themselves into unique adult shapes, and to renew this shape after injury?

Plants and creatures (as creationists delight in telling us) seem to defy a fundamental rule of physics—the second law of

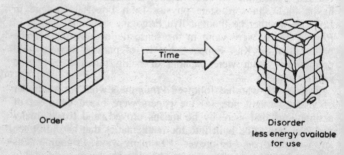

Order

Time

Disorder
less energy available
for use

The second law of thermodynamics states that everything moves to a state of increasing disorder. Do living things defy the law?

thermodynamics. This says that everything tends to degenerate into random heat energy. A burning candle is a vivid illustration. Order (the form of the candle and the structure of the hydrocarbon molecules that make up the paraffin wax) is transformed into disorder (puddles of unburnt wax, much smaller molecules of carbon dioxide gas and water vapor, and randomly distributed heat). Isaac Asimov summed up the second law: "All changes are in the direction of increasing entropy, of increasing disorder, of increasing randomness, of running down."[11]

Life and Nonlife

Yet it is plain that living systems behave differently. They move in the direction of an increase in order, organization and complexity. In addition, each living thing is uniquely capable of reproducing copies of itself. This is true of all organisms, from humans to the simplest unicellular microbes. Each has a form and organization that remains apparently unchanged while its constituent parts are renewing themselves. Thus you and I feel much the same beings today as the day before; yet millions of molecules and cells in our bodies have died and have been replaced in that time. These unique characteristics of living systems—self-organization and self-renewal—have no immediate analogy in the nonliving world.

So what makes us different? What stops us disintegrating? The orthodox scientific explanation has been to say that a living thing is an "open" system—we are open to energy coming in from the outside, and we use it to keep us going. We take out energy by eating, and disperse it by staying alive.

On the other hand there are "closed" systems where no transfer of energy takes place. You might visualize, as an example, the inside of an empty vacuum flask. No heat energy can come from outside. The molecules of nitrogen, oxygen, and so on quickly increase in entropy as soon as you put the sealed lid on, settling themselves into a maximum state of randomness and disorder (in terms of physics, equilibrium). But if you take the lid off and apply heat, the vacuum flask once again becomes an open system. A pattern—convection—would appear among the gas molecules, for heat causes molecules to rise.

Classical science says there is no contradiction between open and closed systems—taken together, the second law still applies.

Thus planet Earth is an open system, the Sun is more nearly a closed system. The Sun gives off huge quantities of energy while increasing its entropy, and some of this is absorbed by the Earth to sustain life here. Overall, there is still a net increase in entropy. To quote Asimov again:

> Life on earth has steadily grown more complex, more versatile, more elaborate, more orderly, over the billions of years of the planet's existence . . . How could that vast increase in order (and therefore that vast decrease in entropy) have taken place? The answer is that it could *not* have taken place without a tremendous source of energy constantly bathing the earth, for it is on that energy that life subsists . . . In the billions of years that it took for the human brain to develop, the increase in entropy that took place in the sun was far greater—far, far greater than the decrease that is represented by the evolution required to develop the human brain.[12]

But while this explanation of energy exchange is true in the mathematical sense, it still begs the most important question. It is not enough for classical science merely to say: "The Sun's energy sustains the evolutionary process." There is still the problem: *how* does the Sun's energy sustain the evolutionary process? *How* does order come from disorder?

As creationist literature points out, the Sun shines on living and nonliving things alike, on human beings and statues of human beings. "It should be self-evident that the mere existence of an open system of some kind, with access to the sun's energy, does not of itself generate growth. The sun's energy may bathe the site of an automobile junk yard for a million years, but it will never cause the rusted, broken parts to grow together again into a functioning automobile. A beaker containing a fluid mixture of hydrochloric acid, water, salt, or any other combination of chemicals may lie exposed to the sun for endless years, but the chemicals will never combine into a living bacterium or any other self-replicating organism."[13]

Laws for Biology

To some of the world's greatest scientists, the problem has seemed so severe as to elude existing laws of physics and chemistry. Erwin Schrödinger, one of the founders of quantum mechanics, said that living organisms might perhaps obey "other laws of physics, hitherto unknown."[14]

Conrad Waddington, too, saw the difficulty, especially as it related to embryos: "A fertilized egg insists, one might say, on changing. The only way to stop it changing is to kill it; that is to say, to alter it from being an open system to a closed one, by preventing the flow of energy, oxygen, and perhaps other raw materials through it. Moreover, in its normal life, as it changes it also very obviously becomes more complex, both in its overall shape and in the number of discernible different parts which can be discriminated within it. For these reasons I do not think any serious embryologists have considered that the second law of thermodynamics can be applied in any simple way to their subject material, in spite of what classical physicists might say."[15]

This, then, is science's greatest gap: the one that Ilya Prigogine has sought to close.

His starting point is the inadequacy of classical thermodynamics to deal with open systems. The second law, he says, is best at describing closed systems where equilibrium has been reached, or nearly reached. Equilibrium, in the sense he uses the word, means a uniform, structureless mixture of heat molecules in random aimless motion—the so-called "heat death" of the Universe, or the air gases inside a vacuum flask.

But open systems (and this includes the whole Universe until it approaches its "ideal" state billions of years hence) are not in equilibrium. They are all, as he puts it "far from equilibrium" —pulsating, fluctuating, and fundamentally unstable. Different laws apply—or at least, additional laws.

Our eyes can deceive us. An open system—a living organism, an automobile, an electricity generator—may look solid and stable. But each is a marvel of nature, precariously balanced, organizing and maintaining itself through a continuous dynamic flow of interlocking forces.

For instance a fountain, seen from a distance, has the appearance of permanency. But on closer inspection, its instability is apparent. Put your hand in a water jet, and you can easily disturb its form. At the same time, like all open systems, it is self-organizing. Take your hand away, and the fountain regains its original shape.

Sudden Change

Prigogine's revolutionary contribution has been to show that as a result of this instability, open structures are continually poised to evolve into something more complex. He has demonstrated mathematically how the fluctuations and pulsations feed back upon themselves, and a critical point can be reached (a *bifurcation*) where there is a sudden shift to a new state.

The change usually happens abruptly and unexpectedly. No matter how gently you open up a tap of water, at a certain point the flow alters surprisingly from a smooth transparent jet (laminar flow) into an agitated cascade (turbulent flow).

Prigogine has also demonstrated that with each new state there is a higher degree of complexity, and therefore, he says, even more potential for change. "The more complex an open system, the more energy it must dissipate to maintain its complexity."[16]

Within the new structure, the self-organizing flux of energy now has extra components interacting among each other, temporarily biding their time and maintaining the appearance of stability until, spontaneously, circumstances combine to shift them into a yet higher form.

"The surprising thing is that quite small fluctuations can result in major changes—and suddenly you find structures of a higher degree of order. Each jump creates new feedback conditions that prepare the way for a new instability enabling the system to evolve further. With new levels of complexity, there are new rules. There is a change in the nature of the 'laws' of nature," he says.[17]

The direct contribution which Prigogine has made to evolutionary theory lies mainly in two densely argued papers published in *Physics Today* in 1972 titled "Thermodynamics of Evolution," in which he speculated on how life could have arisen from nonlife.[18] Citing a number of spectacular chemical phenomena where nonrandom patterns—spirals and other intricate images— can arise spontaneously under certain conditions, he proposed that something of the same sort must have happened to establish the original genetic code.

Indirectly, however, Prigogine's influence spreads far wider than this cautious attempt to explain the emergence of life. His school of colleagues and supporters have eagerly picked up the implications of his work and applied it to every aspect of evolution—social as well as biological.

Self-organizing Universe

At the AAAS symposium, for instance, much was made of Prigogine's suggestion that evolution inevitably meant "novelty and unexpectedness."[19] One speaker said the basic unit of evolution was a "cycle of processes closing upon itself, but always ready to transcend its own organization and to evolve."[20] Another thought evolution could be viewed as "the expression of an inherent playfulness of an always intelligent universe."[21]

The chairman of the symposium, Professor Erich Jantsch of UCAL at Berkeley, is a tireless exponent of the evolutionary vision, which, he says, regards the Universe as a living whole. (He contrasts this with orthodox science, which tries to reduce it to separate bits and pieces that can be measured.) "Life appears no longer as a phenomenon unfolding *in* the universe—the universe itself becomes increasingly alive."[22]

To an outsider such as myself, the symposium had a curiously surreal air, more akin in some ways to missionary fervor than to science as it has lately been practiced. It is all very well talking about the inherent playfulness of the Universe, but it hardly solves the detail of how a bear changed into a whale. Moreover, speakers seemed unanimous in believing that Prigogine offered the *only* alternative to the aridity of orthodox evolutionary thinking—a strikingly similar attitude to that struck by the creationists, who also see themselves as guardians of the only feasible alternative choice.

The result tended to be a succession of speculations and generalities which, while intuitively one may think them true, simply are not testable in the accustomed scientific way. Erich Jantsch not only admits this, but revels in it. In correspondence after the symposium, he chided me for seeking to link cataclysmic events with evolutionary jumps (e.g., a giant meteorite devastating Earth and wiping out the dinosaurs):

The *decisive point* is that self-organizing systems continuously generate their own internal fluctuations, and test their stability, and that it is these internal fluctuations when they become internally reinforced (through autocatalytic and other highly nonlinear mechanisms), that drive the system over an instability threshold to a new structure. There is no stability, period; only meta-stability, or delayed evolution. This is really the crux, and once you have understood this, then you will throw out of your argument Popper and such people for whom something "outside" has to happen, or a theory has to be falsified before it can be replaced. There you are back to the old dualistic adaption syndrome.

No, evolution is *self-transcendent*, that is to say, always ready to reach out beyond the system's own boundaries, *without any need*, just for the joy of it. Only then can evolution be understood as *creative*, and not adaptive. Do you do only what you need to do, determined by outer circumstances and changes? "Man is a crisis-provoker," John Calhoun has said, and so is all evolution. Unless you really fully grasp this, you are not yet with self-organization, but just responses to external happenings—essentially the old equilibrium thinking (including Gould-Eldredge).[23]

Biology and Non-Science

Statements like that in science are breathtaking. They are certainly not yet, as Brian Goodwin believes the new biology will be, "rigorous, scientifically and mathematically." But what Jantsch writes is a powerful affirmation of how profound the challenge to neo-Darwinism is, and Goodwin agrees strongly that the next generation of biologists will have an entirely different concept of the subject.

I don't think biology at the moment is a science at all, at least in the sense that physics and chemistry are sciences. We need to know the universal ordering principles, just as Newton provided them for the inanimate world.

Newton showed us how a moving body under the influence of gravity—a planet affected by the sun, for instance—has a limited set of motions: circle, ellipse, parabola, hyperbola and straight line. Newtonian mechanics gives you those basic restrictions, and then anything is possible—an infinite number of individual variation.

In the living world, something of the same sort must

apply. But we haven't had our Newton to tell us what the laws are—or rather, we know the universal laws of physics and chemistry and mathematics, but we don't have a theory yet that satisfactorily applies them to biology.

The theory may turn out to be electrical, or it may be diffusion-reaction as Prigogine believes, or it may be some other kind of process like that proposed by the physicist Herbert Froelich, where you find long-range cooperative forces of a type similar to a laser—coherent behaviour over large-scale dimensions as a result of the patterns and directions at the atomic and molecular level. It may be field theory, which I have been working on [panel 27]. Or it may borrow something from all these approaches.

But my hope is that the diversity of living forms—or at least their essential features—can be accounted for by a relatively small number of generative rules or laws. It's too soon to start trying to explain how a bear changed into a whale. We need to know the laws that make the basic form possible in the first place.[24]

It is time to go back to the question posed at the start of this survey of dissent: if you reject neo-Darwinism because it doesn't come up with the answers, and if you reject creationism because it is beyond argument, what do you put in their place?

The answer has emerged fairly clearly now, and can perhaps best be illustrated by matching its various elements against what we are normally taught.

New Biology	Neo-Darwinism
Physico-chemical laws govern the creation of biological forms, and are not random in their effect.	Forms are determined by genes, and it is unnecessary to look further than random mutations for the source of variation.
Similarities of form and common structures in the plant and animal kingdoms are basic to understanding evolution.	By measuring the differences between creatures you can deduce their origin and relationship.
Mathematical laws suggest that in the living world, large-scale transformations can happen abruptly.	Evolution is a gradual process, caused by the limited number of favorable gene mutations.

Study of embryonic development will reveal fundamental laws of form and transformation.	Study of variation in adult populations is sufficient to explain evolution.
An organism and its environment are inseparably linked; environmental effects (e.g., maternal, or "Lamarckian") are crucial to evolution, particularly to its speed.	The genes are inviolable and natural selection of their mutations is the sole creative source of evolutionary change; the environment merely weeds out the least fit.
Macro-evolution tends to happen in conjunction with severe environmental upheaval and relaxed population pressure.	Macro-evolution is simply the accumulation of many microevolutionary changes brought about by intense selective pressure on crowded populations.

Now the interesting thing is that when I have showed the two columns to any scientist with an interest in evolutionary processes who is *not a population geneticist*, the one on the left has almost invariably made better sense to them in terms of a strategy for scientific research. The disciplines the scientists came from included mathematics, physics, biochemistry, immunology, zoology, geology, and comparative anatomy.

Yet as we know, the overwhelming weight of research during this century has been in the hands of the population geneticists, and the funding of it continues to be so. They have, as one mathematician said, "cornered the market." I think this blanket of orthodoxy, stifling alternative approaches, is one of the positively harmful effects of a century and more of uncritical devotion to Darwin.

But as I shall try to show in the next section, there has been an even more insidious effect. The efforts to teach the facts of evolution within the straitjacket of a single theory have led repeatedly to fudging and fixing the evidence.

The biggest casualty, all too often, has been the truth.

PANEL 27
What Laws Do Genes Obey?

Brian Goodwin has drawn an interesting comparison:

> Split a bar magnet in two. You find you have two complete fields. The north-south polarity reasserts itself with each magnet, and if you sub divide it the process continues—fields have this quasi-holistic quality.
>
> The parallel with what happens in cell division is very revealing. Cells divide into two in a way which is typical to all organisms, and you can discern a polarity just as you can in bar magnets.[25]

Goodwin stresses that these field processes are fundamental, but that cell division seems to be geometrically ordered only up to a certain stage. "Irregularities tend to accumulate after the fifth cleavage (32 cell stage). We interpret this to mean that the global field which determines the early cleavage planes becomes relatively weaker as cleavage proceeds and other forces emerge and eventually predominate."[26]

Among these forces are genes:

> We think they are modifying influences, not primary instructions. There is no need for them to come into the picture (and no evidence that they do) until several stages of cell division have taken place.
>
> For instance, in some creatures the cleavage starts to take a spiral form after the second or third cleavage. This spiral field is rather more complicated. You have two possibilities—a right-handed or a left-handed spiral. It alternates from cleavage to cleavage up to the seventh, after which the pattern can no longer be followed. This seems to be the point where a gene is able to choose.
>
> It's a bit like pulling the bath plug out and watching the water disappear down the hole in a spiral. You can put your finger in and change the rotation, but the basic pattern will remain the same, and comes from the property of water as it goes down a circular hole. The gene is doing the same thing, selecting one of the possible solutions inherent in the life process.[27]

PART THREE

Dogma

I am the Wilberforce of 1980—but that doesn't actually mean that I'm wrong.

> —Professor John Maynard Smith
> of Sussex University,
> BBC-TV *Monitor*

It is important to guard against fashionable thinking if only because so many people have believed other things in the past equally strongly and have later been shown to be wrong.

> —Peter Andrews,
> head of palaeoanthropology,
> British Museum of Natural History,
> *Guardian*

CHAPTER EIGHT

Monkey Business

Ever since Darwin's work inspired the notion that fossils linking modern man and extinct ancestor would provide the most convincing proof of human evolution, preconceptions have led evidence by the nose in the study of fossil man.

—John Reader,
author of *Missing Links*, 1981

Darwin was circumspect in *The Origin of Species* about how man had emerged from the jungle. Even though the subject was, he admitted in a letter some two years before publication, "the highest and most interesting problem for the naturalist," he said he intended to avoid the whole subject, so surrounded was it by prejudices. In the event, just one sentence referred to the mystery. As a result of his theory of natural selection, he wrote, "light will be thrown on the origin of man and his history." (In later editions he amended this to "much light will be thrown . . .")

In spite of his caution, public opinion quickly grasped the implication of what his theory meant to humankind: man was descended from, or at least closely related to, the apes. In popular usage, *The Origin of Species* became simply "Mr. Darwin's monkey theory," and as such it soon provoked one of the most celebrated and climactic debates science has known.

In June 1860, seven months after the publication of *Origin*, two papers referring to the theory were due to be read at the Oxford meeting of the British Association for the Advancement of Science. Bishop Samuel Wilberforce, widely known as "Soapy Sam" because of his unctuous theological debating technique, decided to use the occasion to affirm the Church's belief in Divine creation, and to ridicule both the idea of evolution and Darwin's explanation of how it happened.

It is one of the minor curiosities of science history that in spite of the eager excitement with which the event was awaited, no detailed note of the proceedings was kept. The library was full to overflowing; one lady fainted from the heat and the drama, according to a newspaper report. Wilberforce spoke for half an hour, ending rhetorically with what he hoped was an unanswerable question, given the prevailing Victorian devotion to the sanctity of motherhood. Although the exact phrasing is not known, what he said was probably:

"If any one were willing to trace his descent through an ape as his grandfather, would he be willing to trace his descent similarly on the side of his grandmother?"

Or in another version:

"Let me ask Mr. Huxley just one question. Is it through his grandfather or his grandmother that he claims descent from a monkey?"

Thomas Huxley, who had attended the debate somewhat reluctantly, not wishing to be "episcopally pounded," as he put it, was stung to a famous reply:

"I will answer your question, my Lord Bishop. An ape may seem to you to be a poor sort of creature, of low intelligence and stooping gait, that grins and chatters as we pass. But I would rather have an ape for an ancestor than a man who is prepared to prostitute his undoubted gift of elegance and culture to the service of prejudice and falsehood."[1]

This single historic encounter changed the climate of opinion in Victorian England almost overnight. Biology had dared to challenge the Church, and biology had won. From now on, Wilberforce remained quiet, and theologians began to find ways of embracing Darwinism instead of fighting it.

Speculation About Man

But in one area at least, the victory was of dubious benefit. In the new atmosphere of unquestioning acceptance of Darwin's theory, the pendulum swung to another extreme of faith. Speculation ran rampant, and nowhere more so than about man's origins. From Huxley's *Man's Place in Nature* in 1863, through Darwin's own *The Descent of Man* in 1871, right up to the stubborn quarrels of modern anthropology chronicled in John Reader's *Missing Links* in 1981, the search for man's past has been littered with vain hopes and invented hominids.

Java Man, Pekin Man, Piltdown Man, Nutcracker Man—all

hang abandoned on the branch of someone's imaginary ancestral tree. It is a cautionary tale for those who believe that, in science, facts always precede theories, or even that scientists are always dedicated followers of fact.

The vogue for reconstructing lifesize models of our ancestors on the basis of the flimsiest fossil evidence (or even no evidence at all) started early. One of the best-known scientists in Germany at the time when Darwinism swept through the academic world was Professor Ernst Haeckel. If Thomas Huxley deserved his contemporary tag "Darwin's bulldog" for the tenacity with which he defended his old friend in public debate, Haeckel can be seen as Darwin's Doberman pinscher, an aggressive and uncritical popularizer on the continent of Europe of everything Darwin stood for.

Haeckel was undoubtedly an interesting, and at times even inspired, biologist. His major contribution to evolutionary biology was his theory of recapitulation. This "great biogenetic law," as he called it, seemed to him to prove evolution beyond doubt. Watching an embryo develop, he said, you could see it passing through the various ancestral forms of life. Thus a human embryo started as a single cell, became wormlike, next showed fish characteristics (e.g., gills), then amphibian ones, then mammalian (e.g., a tail), and finally became a human being.

Although Haeckel's theory fell into disrepute, during the peak years of neo-Darwinist supremacy (panel 28), the revival of interest in embryonic development has led a number of today's biologists to look upon his ideas more favorably.

The trouble is, Haeckel was a rogue.

Frauds and Forgeries

Time and time again, Haeckel doctored his illustrations outrageously to support his biogenetic law. For instance, he wrote:

> When we see that at a certain stage the embryos of man and the ape, the dog and the rabbit, the pig and the sheep, though recognizable as vertebrates, cannot be distinguished from each other, the fact can only be elucidated by assuming a common parentage . . . I have illustrated this signifi-

PANEL 28
Did We Once Have Gills?

The most widely cited evidence of recapitulation in humans concerns the embryonic "gill slits" supposedly showing the fish stage of our ancestry. There are a series of pouches with grooves between them which develop on each side of the neck region in young vertebrate embryos—fish and mammals alike.

But they are not gills as such. These *pharyngeal pouches*, as they are called, serve as guides for the developing blood vessels. In fish, they turn into gills; in mammals, into glands. They cannot be used for breathing, which is the function of gills. They seem, in fact, to be simply an essential and predictable stage of growth common to living embryos before they diverge on their genetically preordained pathways.

The theory of recapitulation is today generally discounted, in spite of it having been dignified as a biogenetic "law." Sir Gavin de Beer wrote: "Seldom has an assertion like that of Haeckel's theory, facile, tidy, and plausible, widely accepted without critical examination, done so much harm to science."[23] Conrad Waddington agreed: "The type of analogical thinking which leads to theories that development is based on the recapitulation of ancestral stages or the like no longer seems at all convincing or even very interesting to biologists."[24] Walter Bock of Columbia said the theory "has been demonstrated to be wrong by numerous subsequent scholars."[25]

cant fact by a juxtaposition of corresponding stages in the development of different vertebrates in my *Natural History of Creation* and in my *Anthropogeny*.[2]

Which indeed he had. But as a matter of biological fact, the embryos of men, apes, dogs, and rabbits are not at all the same, and can easily be distinguished by any competent embryologist. They only *looked* the same, in Haeckel's books, because he had chopped off bits here and there, and added bits elsewhere, to make them seem identical.

Another example was his illustration of the "wormlike" stage through which all vertebrates were supposed to have passed. He published three identical drawings captioned respectively a dog, a chicken, and a tortoise. In 1886, a Swiss professor of zoology and comparative anatomy complained that Haeckel had simply used the same woodcut (of a dog embryo) three times.

Over the years various other forgeries were exposed. To illustrate the "embryo of a Gibbon in the fish-stage," Haeckel used the embryo of a different kind of monkey altogether, and then sliced off those parts of the anatomy inconvenient to his theory, such as arms, legs, heart, navel and other nonfishy appendages. Another time, he altered the shape of embryological drawings to make the braincases of fishes, frogs, tortoises and chickens look the same. Again, he would insert imaginary animals in a neatly graduated progression of forms. On the page, this looked as if it demonstrated life developing from simple to complex—but readers were given no hint that some animals were real, and some pure fiction. An isolated but particularly brazen example of forgery was when he extended the thirty-three vertebrae of a human to thirty-five, and then for good measure tacked on a tail with a further nine.[3]

It says something about the scientific climate of the time that Haeckel remained completely unrepentant about this extraordinary catalogue of falsifications, and that although he was criticized, he was never dismissed or disgraced. In 1908, after a particularly thorough exposure of his methods had received wide publicity, Haeckel wrote airily to a Berlin newspaper:

> To cut short this unsavoury dispute, I begin at once with the contrite confession that a small fraction of my numerous drawings of embryos (perhaps six or eight per cent) are in a sense falsified—all those, namely, for which the present material of observation is so incomplete or insufficient as to compel us, when we come to prepare a continuous chain of the evolutionary stages, to fill up the gaps by hypotheses, and to reconstruct the missing links by comparative syntheses . . . After this compromising confession of "forgery" I should be obliged to consider myself "condemned and annihilated" if I had not the consolation of seeing side by side with me in the prisoner's dock hundreds of fellow-culprits, among them many of the most trusted observers and esteemed biologists. The great majority of all the diagrams in the best biological textbooks, treatises and journals would

incur in the same degree the charge of "forgery," for all of them are inexact, and are more or less doctored, schematized and constructed.[4]

One of Haeckel's invented embryos showed the head of an ape on the body of a man, and for fossil-hunters the world over, the hoped-for discovery of a true grown version of this hybrid creature was the supreme prize: the elusive missing link.

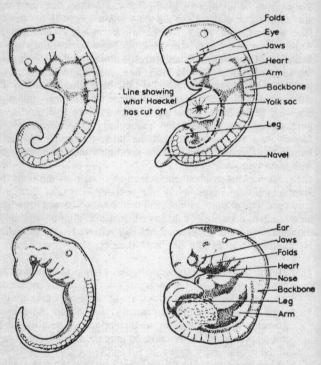

Two examples of Haeckel's alterations to embryonic drawings. Above is his illustration showing "embryo of a Gibbon in the fish-stage." Excised organs not suiting his purpose including jaws, arms, legs and other mammalian features. Below is his "embryo of man in the fish-stage." The textbook original (right) has had more than half of its essential organs removed or mutilated.

Invented Apemen

Thus far, the only fossil specimens of early man which had been found were Neanderthals. To Victorians, for a variety of reasons, these didn't count. Their heads were too large to fit neatly on the gradually ascending scale of brain size which was supposed to lead from apes to men. The rock deposits from which they were dug up were, at that time, not possible to date accurately. So the skeletons were dismissed as human oddities—fossilized idiots perhaps, or, in Huxley's opinion, the remains of a rickety Cossack from the Napoleonic wars.[5]

What Haeckel wanted, and what Darwinism needed, was something genuinely halfway up the ladder. Widely though the theory was accepted in regard to nature in general, there was a lingering suspicion among Victorians that man was somehow superior to the law of natural selection. The absence of fossils demonstrating an intermediate was an embarrassment. Haeckel, absolutely certain that such a creature must have existed, commissioned a drawing of it, and gave it a resounding Latin name: *Pithecanthropus alanthus*, the apeman without speech.

And sure enough, in slightly mysterious and perhaps dubious circumstances (panel 29), bones were found. A former pupil of Haeckel's at Jena University, a young Dutch doctor named Eugene Dubois, left for the Far East in 1887 determined to discover the first man. After three seasons of almost fruitless searching, he achieved his ambition in the autumn of 1891, at Trinil on the island of Java. In September, he discovered a large upper right molar tooth; in October, the broken cap of a skull which was indubitably smaller than modern man's. A year later, also at Trinil but fourteen meters away from the first finds, he came across a human leg bone, almost complete, plus several more leg bone fragments, and another molar tooth.

It was little enough to go on (indeed, Dubois at first thought the skull cap and teeth belonged to a chimpanzee), but he wrote at once to Haeckel asking if any of this could be considered the remains of early man. Haeckel was in no doubt. He telegraphed back immediately: "From the inventor of Pithecanthropus to his happy discoverer."

From then on, both men were sure they were on to the real

PANEL 29
How Genuine Is Java Man?

Early commentators on the Java discoveries could see no reason for associating the broken skull cap and the two molar teeth with the leg bone fourteen meters away. Dubois suggested that his apeman might have been eaten by crocodiles, and the bits of bone washed into their separate places by the river, to which a critic retorted that it must have been an exceptionally well-harnessed river to have held its course for half a million years.[26]

Haeckel's former professor, Rudolph Virchow, refused to chair the meeting when Dubois came to Berlin in 1895 to exhibit his skull cap and thigh bone. Like Dubois himself when he had first dug it up, he thought the deep suture in the skull exactly like that of an extinct ape of an unknown kind. "In my opinion this creature was an animal, a giant gibbon, in fact. The thigh bone has not the slightest connection with the skull."[27]

Dubois, under attack, became secretive and shifty about his fossil finds. On his voyage home he brought 215 cases of fossils, although the sum total of human remains consisted of the partial skull cap, the leg bone plus a few fragments of leg bone, and two teeth. Persuaded by Haeckel, he had come to believe the skull cap to be transitional. He estimated a brain capacity of 900 cubic centimeters, notoriously difficult though it is to arrive at such an exact estimate on the basis of a portion of the cap of a skull.

What he did not divulge were the two fossilized human skulls he had found at Wadjak the previous year, at approximately the same geological level. Had he done so, his many critics would have undoubtedly pointed out that this made it even more likely that the fossil leg bone was from a normal, recent human being. He also concealed, for reasons best known to himself, the fragments of other leg bones, and one of the teeth, hiding these and other fossils beneath his dining room floor.

It was not until the 1920s and 1930s that Dubois spasmodically announced these finds, usually under pressure from other palaeontologists. One who visited him, G. H. R. von Koenigswald, who was about to leave for an expedition to Java to see what more could be discovered, found the fossil collection in poor condition after forty years, with some of the labels identifying the exhibits lost.

The result has been continuing doubt about Dubois's work. A 1906 expedition to his site at Trinil excavated 10,000 cubic meters of soil down to a depth of twelve meters, without producing a single hominid fossil. Such slender evidence as there was—a human tooth, traces of hearth foundations and charcoal—suggested that modern man had been a contemporary of *Pithecanthropus*, if indeed *Pithecanthropus* was anything more than a large gibbon.

Subsequently, during the 1930s von Koenigswald recovered skull fragments from a site forty miles away from Trinil. They are similar to the skull cap found by Dubois, but smaller, making a reconstruction even more difficult. Other finds were made in 1953 and 1961, and today the consensus is that Dubois's bits and pieces of apemen once belonged to a group of human predecessors known as *Homo erectus*, who lived between 800,000 and 200,000 years ago, and whose remains have been found sparsely in a number of other parts of the world. The authenticity of the Java

bone fragments is supported by a common level of fluorine content, consistent with a date of *c*. 500,000 B.C.; potassium argon dating of material found at the same level indicates a date some 200,000 years earlier. So the fossils have come to be accepted, somewhat doubtfully, as genuine examples of an early relative of man.

However, many problems still remain, both about Java Man and about *Homo erectus* in general. The fossil-bearing layers in Java are extremely difficult to date accurately, and to relate to one another, because of volcanic activity over the years. The severity of the upheavals was demonstrated as recently as 1909, when a disastrous flood followed an eruption, fourteen inches of rain fell in a day, more than 500 people died, and whole villages were swept away and buried beneath ash and mud. This kind of local catastrophe illustrates vividly the difficulty of being sure that Trinil level which Dubois excavated is the same geological age as those investigated by von Koenigswald forty miles away. A number of von Koenigswald's skull fragments came from a level so deep that they may well be two million years old.

thing. In 1849, Dubois claimed scientific recognition for his find with a paper entitled *Pithecanthropus erectus, a Human-like Transitional Form from Java*. Haeckel commissioned a life-size model and exhibited Java Man in museums throughout Europe. Darwin was vindicated no less dramatically than when Huxley triumphed over Soapy Sam.

Yet as G. K. Chesterton pointed out at the time, "people talked of Pithecanthropus as of Pitt or Fox or Napoleon. Popular histories published portraits of him like the portraits of Charles I or George IV. A detailed drawing was reproduced carefully shaded to show the very hairs of his head were all numbered. No uninformed person, looking at its carefully lined face, would imagine for a moment that this was the portrait of a thigh bone, of a few teeth, and fragment of a cranium."[6]

Pekin Man

Java Man's successor was Pekin Man, around whose discovery and reconstruction lies a similarly ambiguous tale. In 1929 an almost complete skull cap, very similar to those in Java and described as "apelike," was found in an infilled limestone cave at Choukoutien, near Pekin. From then until the outbreak of World War II the site was continuously excavated, yielding fragments of some fourteen skulls, twelve lower jaws and 147 teeth, thought to represent about forty individuals. Except for scraps of limb bones, no other primitive apeman remains were found, although several skeletons of modern men were found at a higher level.

Using bits of bone from various parts of the site (the jawbone came from a level more than twenty-five meters higher than the skull and facebones), a complete skull was reconstructed. A Mrs. Lucille Swann, a sculptress then resident in Pekin, modeled a woman's features on a cast of the skull, and the result was dubbed "Nellie," in which form she has appeared in many textbooks, complete with hair, a sagacious expression, and an improbably thick neck.

How good a likeness this is, we shall never know. Early in World War II, at the beginning of the Japanese occupation of China, all the skulls were lost, perhaps in the sinking of a ship while they were being transferred to safety, perhaps through some other mishap. There have been persistent rumors that the

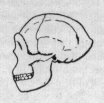

Three stages in the emergence of "Nellie," the reconstructed face of Pekin Man used in the trial of John Scopes. First, the skull cap is assembled from assorted fragments (left); next, an entire skull is worked out, without using any further fossils (center); finally, a sculptress resident in Pekin fleshes out the skull to make it look nearly human.

fossils were brought back to America after the war, where they conveniently disappeared because too close an examination, it was suspected, would reveal them to be more apelike than Nellie appears, and certainly too primitive to have been using the quite sophisticated stone tools found alongside them.

From this has stemmed another renegade theory, that (as was also suggested by the 1908 expedition to Java) relatively modern men were living alongside this creature, whatever it was; and at Choukoutien they were eating its brains as a delicacy, in much the same way as some Chinese do today. It is an attractive speculation, not least because it solves at one stroke the puzzle of why no other parts of the skeleton are found in the cave, why the skulls are comprehensively crushed, why the tools are the same as those used by early man around 25,000 B.C., and why the limestone "cave," as it is always described, is more in the nature of an artificial quarry.

Nebraska Man

However, it is not the orthodox view, and if it came to be accepted Nellie would have to go the way of the most audacious reconstruction yet made, that of Nebraska Man. In the early 1920s, before the Tennessee "monkey trial" of the school-teacher John Scopes (see below), the teaching of Darwinist evolutionary theory in U.S. schools was even more an educational issue than it is today. William Jennings Bryan, a Nebraskan politician, was vigorously contesting through the courts the idea of children being taught they were descended from apes. In 1922, H. F. Osborn, head of the American Museum of Natural History, received a fossil tooth found in Pliocene deposits (i.e., at least two million years old) in Bryan's home state.

Osborn was overjoyed. The tooth, he decided, was definitely humanlike. There must have lived, in those times, Nebraska Man—or to give him a dignified Latin tag, *Hesperopithecus*. He exulted: "The Earth spoke to Bryan from his own State of Nebraska. The Hesperopithecus tooth is like the still, small voice. Its sound is by no means easy to hear. But this little tooth speaks volumes of truth, in that it affords evidence of man's descent from the ape."[7]

He was fully supported in England by the most eminent of anatomists. Professor Sir Grafton Elliot Smith of Manchester authorized a detailed reconstruction of Mr. and Mrs. *Hesperopithecus* that duly appeared in the *Illustrated London News*, complete with a vivid background showing horses and camels in exotic prehistoric surroundings.

Alas, the tooth was later identified as belonging to an extinct pig, and today no more is heard of Nebraska Man. It is one of the ironies of history that he nevertheless served his purpose in another famous Darwinist triumph, the Tennessee "monkey trial" of 1925. When John Scopes was accused of teaching Darwin's theory contrary to Tennessee state law, the *Hesperopithecus* tooth was proudly displayed as evidence that man had a long evolutionary past; so were Eugene Dubois's fossils, even though Dubois himself had ceased to believe in them; and so was Piltdown Man, now known to be the most notorious forgery of all, consisting of a cleverly doctored ape's jaw that precisely fitted a modern human skull cap, alongside ancient fossil teeth.

So the trial that became a turning point in U.S. educational history, not to be significantly challenged for the next half-century, was steered toward its verdict by a pig tooth, two dubious fossils subsequently repudiated by their finder, and an outright fake exhibit whose perpetrator is still not known.

"I suspect that unconscious or dimly perceived finagling, doctoring, and massaging are rampant, endemic, and unavoidable in a profession that awards status and power for clean and unambiguous discovery," Stephen Gould said about scientific research in general.[8] So perhaps this tale of Darwinian skulduggery is not so extraordinary as it first strikes an outsider as being. Piltdown Man and Haeckel's embryonic apeman apart (although these are pretty hefty exceptions), there is nothing in this catalogue of falsity which is totally and deliberately fraudulent.

More, the lesson is that insatiable public curiosity about man's origins on the one hand, and deeply held scientific prejudices together with fragmentary and ambiguous fossil evidence on the other, combine to give each generation the ancestor it wants, and each palaeontologist the fossils he seeks. Thus Eugene Dubois could go out to Java, find a human leg bone and an apelike skull fifteen whole paces from one another, buried in dust and mud that had accumulated over hundreds of thousands (perhaps millions) of years, and decide cheerfully that they belonged to the same creature.

Piltdown Man

Thus, too, the uncritical acceptance over so many years of Piltdown Man, uncovered in a Sussex gravel pit in 1912. Here, the fashionable prejudice that gave the skull its ready appeal was the prevailing obsession with brain size. It was generally thought that intelligence was directly related to the amount of gray matter we had, and that man could only have achieved his special place in the animal world because he had the mind to do so. The human cranium of the Piltdown specimen seemed to confirm the idea.

We know now that this was wrong. But the story of Piltdown is only an extreme example of the constantly changing versions of man's past. We have had Man the Toolmaker, when anthropologists believed that mastery of stone implements enabled our ancestors, by a kind of Darwinian survival of the cleverest, to rise above their ape relatives. Then we had Man the Killer-Ape, otherwise known as the hunting hypothesis, in which the stone tools were supposed to give early man an advantage over other creatures through their use as weapons. Today, that theory has been discarded and replaced by a preoccupation with Man the Upright Ape: it turns out that man learned to walk before he could think. The young U.S. fossil-hunter Don Johanson's stunning discovery of "Lucy," a forty-percent-complete skeleton of an ape-woman who died in her late twenties about 3.5 million years ago, shows this conclusively. Her brain was as small as an ape's but she walked upright. Why? What possible Darwinian advantage could this confer? All the other monkeys and apes around at the time—and there were many more types then than now—could surely have outrun Lucy and her kind with their speedy, four-footed gait.

It is easy enough to understand why fashions change. Major finds happen seldom, and when they do there is a natural tendency for their discoverers to exaggerate and dramatize their importance, and at the same time use the opportunity to elaborate a new version of man's past. John Reader thinks that on the basis of the available fossil evidence, the study of fossil man hardly deserves to be more than a subdiscipline of palaeontology or anthropology (actually, it calls itself palaeoanthropology): "the entire hominid collection known today would barely cover a billiard table."[9] Moreover, such fossils as are found are usually so fragmentary (Lucy is a rare exception), and in such ambiguous surroundings, that many interpretations are possible.

What is perhaps harder for an outsider to understand, except in the context of a declining faith in Darwinism, are the passions and jealousies which the subject arouses. These came to a head at the end of 1980 and the beginning of 1981 in an outstanding display of academic ill-temper and prejudice conducted in the correspondence and leader columns of *Nature*. It was sparked off by an apparently innocuous exhibition within the British Museum of Natural History, directed mainly at schoolchildren, entitled "Man's Place in Evolution." But as the debate widened to include contributions from many of the evolutionary spokespeople we have met in this book, it became clear that what was on trial was Darwinism itself.

Man's Jerky Past

Such evidence as we have of man's past presents the familiar challenges to Darwinian theory: in particular, to gradualism and the idea that natural selection alone is responsible for change. It had always been assumed that as brains steadily grew larger and man's ancestors became cleverer, so we learned to walk upright and developed the use of tools. But as Lucy shows, this is simply not the case.

We learned to walk (or rather, *some* kind of apelike creature stood up and walked) at least a million years before there is any evidence of large brains or handmade tools. Brain size does not follow a Darwinian progression. In humans it ranges widely from 1,000 to 2,000 cc, and has no correlation with intelligence. The two largest brain capacities yet recorded, 2,800 cc, belonged respectively to a U.S. senator and to an idiot. Australian aboriginals today have brains approximately the same size as one of our "primitive" predecessors, *Homo erectus* (Dubois's Java Man has been placed in this category). Our immediate forerunners, the Cro-Magnon people of 50,000 years ago, actually had brains on average slightly larger than ours.

But if brains haven't grown in a convenient, gradual, Darwinian way, nor have the other characteristics which mark out man from his fellow creatures. Instead, they appear in the fossil record with the same suddenness as so many other evolutionary innovations—the "herky-jerky" pattern established by Niles Eldredge and Stephen Gould.

Take the change to upright walking. Very considerable anatomical changes divide us from apes—not so large as a comparable earlier change, the transition from fish fins to amphibian legs, but substantial none the less. They involve principally the feet, pelvis, vertebral column, and related muscular systems. The assumption had always been that a period of perhaps ten million years was needed for them to take place.

But the work of Allan Wilson and Vincent Sarich, the Berkeley immunologists, has disproved this. They developed a method for finding out when various species separated from each other, by consulting what they call a "protein clock." To make the clock work, they count up how much of a creature's genetic code is identical to another creature's; the bigger the difference, the further apart in time they separated. When they found, to their surprise, that the genes in humans and in African apes differ by less than one percent, their clock told them that we split up only about five million years ago; at most, six million. But Lucy was walking, and so were other hominids elsewhere in Africa, only about one million years later.

Cladism in Dispute

The method chosen by the British Museum of Natural History to accommodate this sort of uncomfortable finding in its exhibition would seem scientifically impeccable, and at the same time very adroit. They have stuck to such facts as are known, and left the theories of origins as far as possible on one side. And this, of course, is what has made traditional Darwinists so angry. When the oldest and perhaps the most prestigious natural history museum in the world starts beating a retreat from Darwinism, a funeral march is on the way.

The innocent-seeming device by which the museum has drawn away from the traditional evolutionary picture, with its forest of ancestral family trees, goes by the name of *cladistics* or cladism. It is a relatively new system of classifying living things, and, say its supporters, it has two great merits: it is based on fact rather than opinion; and it can be tested. Further, at heart it is so simple that a child can understand it.

The "facts" on which it is based are observable features which groups of living things can be seen to have in common. All mammals, for instance, are vertebrates with fur or hair, mammary glands, and three separate bones in the ear. We humans have this too, and so count as mammals: we are more closely related to dogs and cats than to fish or frogs. By identifying the maximum number of features we share with particular mammals, we can establish that our closest living relatives are chimpanzees and gorillas.

Among our fossil predecessors, we share most features with the Cro-Magnon people, and are therefore most closely related to them. We share less characteristics with Neanderthals, and are therefore a little more distantly related to them. And so on back to the earliest apelike creature, where there is just enough evidence of such features as a shortened muzzle to make them closer relatives than chimps or gorillas. Cladistics, in short, relates animals to one another by measurable likenesses.

That children enjoy making sense of nature's diversity through this system is obvious from their reaction to the two displays at the museum organized on this basis (the other being about dinosaurs and their place in the world). They are well acquainted with the principle, since the Venn diagrams and "sets" of new mathematics that they are taught at school use it all the time. As a matter of policy, the museum decided a few years ago to make its exhibits understandable to the broadest spectrum of visitors, and to design them with a fifteen-year-old "target" audience in mind. In this they seem to have succeeded admirably.

Why, then, the heat? For as Colin Patterson, a senior palaeontologist at the museum who embraces cladistics wholeheartedly, has written: "Whenever and wherever cladistics is

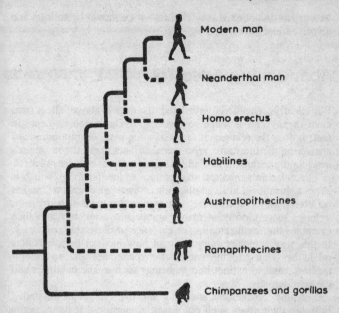

Modern man

Neanderthal man

Homo erectus

Habilines

Australopithecines

Ramapithecines

Chimpanzees and gorillas

A cladogram of man's relationship to fossil apemen. Superficially like the traditional tree of life, it simply records the number of shared characteristics that we have with our predecessors, and makes no attempt to build an ancestral relationship.

discussed, there is controversy. The virulence of this controversy and the partisan feelings evoked are remarkable in so old and apparently harmless a discipline as systematics; there has hardly been such bad temper in biology since the violent arguments provoked 120 years ago by Darwin's theory of evolution by natural selection."[10]

There are some technical objections to cladism which carry a certain weight (e.g., how can you know you have identified all the relevant features?). It also has a large nuisance quotient in reclassifying all manner of creatures that, in textbooks and other museums, are currently displayed alongside one another. Thus coelacanths and lung-fish, because of their four limblike appendages, are closer cladistic relatives to humans than to fish. Birds appear closer to mammals than to reptiles. Indeed, reptiles as a group do not exist, say cladists—they should be split up and dispersed elsewhere.

Abandoning Ancestry

But clearly, what has infuriated its critics above all is that cladism pays no homage to theories of Darwinian descent. In fact, it does the reverse. If understanding man's evolution means answering the question, Who were our ancestors? the museum's newfound attachment to cladism specifically avoids the issue.

The museum's director, in a preface to the handbook which is almost sensational in its implications, says the exhibition "makes no attempt to reconstruct the history of human evolution. Instead, it looks more closely at man's unique characteristics and then examines the fossil remains for evidence of these characteristics. In this way a picture is built up of how modern human beings might be related to the ancient ramapithecines and australopithecines, and to extinct human beings such as the habilines and the neanderthals."[11]

Note the "makes no attempt" and the "might be related." Between them, they spell out a tacit admission that the museum is refusing to become involved in Darwinian speculations about the descent of man. The exhibition breaks abruptly with all past tradition, where a tree of ancestry has shown each successive new form of ape man related by descent to a previous ape man. Cladists deny that this can ever be established. "We assume that none of the species we are considering is the ancestor of any of the others. This is a fairly safe assumption, because so many animals have lived and died during the history of life on Earth that the chances of finding—and recognizing—any particular ancestral species are very, very small."[12]

This is not so much a debate as a confrontation. Darwinists assume the opposite. Random gene mutations have led gradually to modern man, therefore we *must* be descended from at least some of the intermediate apemen that have been discovered. Beverly Halstead, Reader in Zoology and Geology at Reading University, took up the challenge on behalf of the traditionalists:

> With regard to both dinosaurs and fossil man, it is evident
> that the application of cladistics is quite inappropriate. The
> well attested sequence of human fossils representing sam-

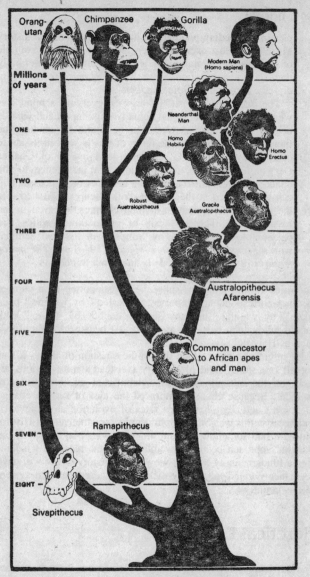

Man's ancestry, according to the current thinking of fossil experts.
The faces, however, are "pure fiction."

ples of succeeding populations has, until the Natural History Museum's latest exercise, been taken as a classic example of the gradual evolution of a single gene pool. Certainly there is not any serious doubt about *Homo erectus* being directly ancestral to *Homo sapiens* . . .

The questions that should arise in everyone's mind are: what is this all about, what actually is going on and what is behind it all? The answers can be found by reading the literature of cladistics. The tenor of this is seen in its abuse of E. Mayr and G. G. Simpson and indeed of Charles Darwin himself, because of their firm adherence to the concept of gradualism and to the idea that the processes that can be observed at the present day, when extrapolated into the past, are sufficient to explain changes observed in the fossil record. The synthesis of population genetics and palaeontology presented by Simpson in his two seminal works *Tempo and mode in evolution* (1944) and *The major features of evolution* (1953) is anathema to cladists.[13]

To which Colin Patterson, subsequently supported by a number of other correspondents, reported: "attested" by whom? "been taken" by whom? "not any serious doubt" by whom? Why should gradualism be taken as a fact just because "authorities" such as Halstead say so?[14]

He might also have added that in the question of abuse, it was not all one-sided. One of Beverly Halstead's most curious attacks on the museum was not scientific at all, but stridently political. Because cladism embraced the idea of sudden change (it doesn't actually—it regards rates of evolution as irrelevant), and because this was Stalin's philosophy too, the Natural History Museum had somehow become infected with Marxism. "If the cladistic approach becomes established as the received wisdom, then a fundamentally Marxist view of the history of life will have been incorporated into a key element of the educational system of this country."[15]

Heretical Exhibitions

The unfriendly editorial in *Nature* which summed up the debate found this particular charge silly—but it regarded the museum's retreat from Darwinism as a most serious matter. "Darwin's Death in South Kensington," ran the headline. The article said the charge that the museum's new exhibitions were "shot through with heresy" needed to be answered. If the museum had serious doubts about the theory of evolution, "the rot at the museum has gone further than Halstead ever thought. Can it be that the managers of a museum which is the nearest thing to a citadel of Darwinism have lost their nerve, not to mention their good sense?"[16]

To me, it seems that what the row emphasized is the desire among a growing number of evolutionary specialists to go back to the drawing board: to the fundamentals of patterns and forms of nature. We saw it emerging in the previous chapter, in the search for underlying rules of embryological development, for the mathematics of changing shapes, and for the biological laws of structure. Cladism is an aspect of this. It gives museum scientists breathing space—a chance to stand away from the doubts about evolutionary theory and look instead at basics.

Peter Forey, a palaeontologist at the museum, told me: "We have come to believe that potentially observable characters are all important, rather than some theoretical speculations about the continuity of gene flow."[17] Norman Platnick, of the American Museum of Natural History in New York, believes "we are hardly likely to achieve any understanding of the evolutionary process until we have achieved an understanding of the pattern produced by that process, and we have hardly begun to understand the pattern."[18]

Perhaps the most striking evidence of how swiftly and profoundly this new attitude to biology is taking root comes from Colin Patterson. It was he who wrote, in 1978, the museum's definitive book *Evolution*. For all its cautious reservations, it is primarily a statement of the modern synthesis. "Today's theory of evolution is unlikely to be the whole truth. Yet today's neo-Darwinian theory, with all its faults, is still the best that we have. It is a fruitful theory, a stimulus to thought and research, and we should accept it until someone thinks of a better one."[19]

Two years later, in 1980, he had moved a long way from that position: "In my view, the most important outcome of cladistics is that a simple, even naïve method of discovering the groups of systematics—what used to be called the natural system—has led some of us to realize that much of today's explanation of nature, in terms of neo-Darwinism, or the synthetic theory, may be empty rhetoric.

"If this is the result of cladistics, to give neo-Darwinism a good shake, then perhaps its critics are right to get excited."[20]

*   *   *

So where does this leave the search for man's origins? Not all the museum experts are as ruthless in excluding certain areas of investigation and speculation as the cladists. Peter Andrews, head of the palaeoanthropology section, has put his name to a traditional (more or less) ancestral tree. Lucy appears under her official title *Australopithecus afarensis*. Next come two other sorts of australopithecines and *Homo habilis*, the fruits of the many years spent in Kenya by the Leakey family. These seem to have died out and been replaced by *Homo erectus*, among whom Dubois's Java Man and Nellie from Pekin take their place. Finally, just as mysteriously and abruptly, came the arrival of Neanderthals, Cro-Magnon people, and us.

It is certainly more complicated than the straight-line-of-descent people would have us believe—more a bush than a ladder, as Stephen Gould has put it.[21] The biggest change in thinking since Darwin's time has been the abandonment of a search for the Missing Link between ourselves and apes. Instead, the hunt is on for the Common Ancestor. According to Peter Andrews, we already know quite a lot about him/her:

> The common ancestor between the African apes and man was probably a medium sized (10–30 kg) active primate with grasping hands and feet and mobile arms and legs (revealing its tree-living past), but its arms would not have been lengthened, so that it would not have habitually swung its arms in the trees as do some of the living apes. Its body was probably covered with hair, with a broad chest and lacking a tail. Looking at its rather short and flattened face, its eyes would have been widely set over a squashed-looking upturned nose. Its strongly built jaws had four or five cusped upper and lower molar teeth, projecting canines and broad incisors.
>
> This ancestral hominid would have been highly social, probably with several females and males making up the group, and the males would have been considerably larger than the females. It inhabited the tropics, almost certainly in Africa, and probably lived in rather open woodland or wooded savanna habitats. It would have eaten a variety of vegetable foods, including nuts and fruit for which it would climb trees, and seeds, roots and fungi which it would have eaten on the ground. Some of these characters were retained or changed in one direction during the subsequent evolution of the African apes, and others were retained or changed in another direction during the evolution of man.[22]

*   *   *

All this is inference from many scientific disciplines, without a hint of fossil evidence to support it directly. Nor is there ever likely to be. I asked Peter Andrews how many of these creatures were thought to have lived—thousands? hundreds of thousands? millions?

"Oh, a fairly small number over a fairly short period of time, if you believe the new ideas about evolution."

The new ideas meaning a more sudden origin of species than neo-Darwinism suggests?

"That's right."

So the chances of finding fossils of this ancestor of ours are pretty remote?

"Vanishingly small, I would say."

What about the depictions of this creature—how accurate are they?

"Pure fiction in most respects."

And the models in "Man's Place in Evolution"?

"Much the same. Anatomically we can probably get fairly close, but all the rest is sheer invention. Skin colour, how much hair—we can't possibly know. In many ways I wish we didn't have to flesh these creatures out."

Nature, I think, got it wrong. The museum, far from losing its nerve or its good sense, seems to be facing up to our ignorance about man's origins with a most creditable frankness and objectivity.

But what of Darwin himself? How does his reputation stand? A larger-than-life statue has him sitting broodily in an armchair gazing at an exhibition that celebrates the museum's centenary, and explains its aims. Symbolically perhaps, or tactfully, the contentious cladist displays are out of his sight, the dinosaurs in the great hall immediately behind him, the fictional models of fossil man upstairs above his right shoulder. Darwin died just a year after the museum originally opened, and ever since it has truly been, as *Nature* called it, a citadel of Darwinism. Can it go on being one?

CHAPTER NINE

Darwin's Legacy

> Then, with a suddenness only less surprising than its completeness, the end came; the fountains of this great deep were broken up by the power of one man and never in the history of thought has a change been effected of a comparable order of magnitude.
>
> —G. J. Romanes,
> *Nature*, 1881

Nineteenth-century natural history needed a Darwin. In the physical sciences, researchers were striding ahead in their attempts to find laws that united the phenomena they were studying. Faraday showed how the many different forms of energy—heat, light, electricity, and so on—were subject to the same set of rules. Organic chemistry opened up a way of determining exact arrangements among atoms. Physicists were elaborating the molecular theory of gases with astonishing success, were unifying electricity and magnetism, and laying the foundations of thermodynamics. Astronomers gazed into the skies and, following Newton's unchallengeable laws, saw that God's part in the arrangement of the heavenly bodies must have been to set the mechanism in motion, and then to have ceased interfering.

But when it came to the mystery of life on Earth, and man's apparently preordained place at its summit, different standards of judgment applied. Here the theologians ruled, insisting that every word of the Biblical creation as described in Genesis was true; and many of them were still preaching that the date for creation was the one worked out by James Ussher, Archbishop of Armagh, two centuries before—that is, 4004 B.C. at nine o'clock in the morning.

"Shall Biology alone remain out of harmony with her sister sciences?" lamented Thomas Huxley. But if the nineteenth century needed a Darwin, Darwin himself was indebted no less to those scientists who had preceded him and prepared the way.

To geologists particularly he owed much. When he started his voyage of discovery on H.M.S. *Beagle*, geology was in the process of transforming itself. Until then, the prevailing belief was in the catastrophism of, for instance, Baron Cuvier. Most of the 600 members of the Geological Society thought that the Earth was relatively young, and that it had undergone a series of cataclysmic upheavals and extinctions followed by the development of new life forms, supernaturally created. But in 1830 came the publication of a book that challenged this: *Principles of Geology*, by Charles Lyell.

Lyell was little known in geological circles, although already, aged thirty-three, a Fellow of the Royal Society. Trained as a lawyer, he brought his legal mind to bear on the field research carried out by numerous specimen-hunters during the previous forty years. The subtitle to his book, which eventually ran to three volumes, stated boldly his theme:

An attempt to explain the former changes of the Earth's surface, by reference to causes now in operation.

Lyell believed passionately that there was no need to invoke a supernatural agency for events that had happened in the past. Now that physicists, chemists and mathematicians were showing that universal laws governed nature, surely, he said, the same must apply to the formation of Earth. Given time enough, the gradual processes of erosion and occasional volcanic activity could amply explain our present habitat.

Thus uniformitarianism was born. In essence, it meant that the forces of nature were unalterable, and could be measured. "The present is the key to the past," he wrote. There does not seem to have been a particular dramatic moment when Lyell's views won the day, and became accepted as the new orthodoxy. Rather, it was a question of successive publications and written arguments by Lyell gradually persuading already confused geologists that his case was scientifically more rational.

At all events, within a decade of his book's publication, the date of Noah's Flood had vanished from the arena of scientific debate. So far as the young Charles Darwin was concerned, Lyell had made him a vital gift: a long enough timespan for life on Earth (like Earth itself) to have evolved gradually, step by infinitesimal step. Later in life he was to write: "I feel as if my books came half out of Sir Charles Lyell's brain." When he stepped on board H.M.S. *Beagle* in 1831 he took with him the first volume of *Principles of Geology*; the second reached him in South America. On his return, it was to Lyell he went for advice and friendship—"I saw more of Lyell than of any other man, both before and after my marriage."

Predecessors

So Lyell's precious contribution to Darwin's understanding is well recognized. Instead of a world 6,000 years old, he and his fellow geologists were now talking in terms of millions of years—perhaps hundreds of millions. What is perhaps less well known is that biology also was in a state of some ferment. It is one of the less pleasing sides of Darwin's otherwise affable and scholarly nature that he could never bring himself to acknowledge a debt to the many predecessors in his field who were puzzling about the origin of species.

There was, for instance, his grandfather Erasmus Darwin, who died seven years before Charles was born. That Charles had studied the old man's work is evident. His first tentative draft paper leading eventually to *The Origin of Species* was headed *Zoonomia*, the title of his grandfather's most famous book, published in two volumes in 1794 and 1796. These formed part of an extraordinary collection of writings, both in prose and in verse, published about the same time, in which Erasmus put forward his ideas on evolution.

The foremost medical practitioner of his time (so successful that he was able to turn down an invitation from George III to become the Royal physician), poet, philosopher and sage, he was the first person to suggest with authoritative evidence that all things were descended from a common ancestor. Like Lyell's predecessor, James Hutton, he presumed a very old Earth—"millions of ages," as he put it. He became one of Europe's most celebrated personalities. *Zoonomia* was soon translated into French, German, Russian and Italian; a pirated edition appeared simultaneously in New York.

With hindsight, we can see that Erasmus Darwin's seemingly bizarre writings (when else has poetry been regarded as the proper language of science?) anticipated almost every important evolutionary idea that was subsequently put forward. He saw that animals competed with each other—as did plants—and that this was a means of change. He saw that sickness could be inherited, and this could change things too. He saw that the competitive male sexual pursuit of females was another element of evolutionary change. And he anticipated Lamarck (and Darwin, too, late in life) in suggesting that acquired characteristics could in some degree be passed on.

Lamarck, too, had a major contribution to make. However, in the end, history judges the central part of his theory, he deserved far more credit than he ever received for observing that evolution had taken place as a result of successive variations. Yet all Darwin could find to say about his grandfather was: "It is probable that hearing rather early in life such views maintained and praised may have favoured my upholding them in a different form in my *Origin of Species*"; and Lamarck he casually dismissed as having "mistaken ideas." There is a certain ungraciousness about this which prompted Professor C. D. Darlington of Oxford to complain of Darwin: "he damned Lamarck and also his grandfather for being very ill-dressed fellows at the very moment that he was engaged on stealing their clothes."[1]

Contemporary Evolutionists

There were other unacknowledged influences on Charles Darwin as he grew up. Increasingly, the origin and growth of life, particularly of man, became a subject for afternoon reading and dinner table conversation. When he was ten, in 1819, a well-known doctor, William Lawrence, published *Natural History of Man*. It was a fascinating book, for his central theme was that racial differences had arisen through heredity, in much the same way as you can see different characteristics in a litter of kittens; and here he foreshadowed Mendel's genetic discoveries that were to become the basis of neo-Darwinism a century later. But this remarkable insight went almost unnoticed in the chorus of denunciation that greeted what he had written.

For the truly scandalous appeal of the book was the message, explicitly spelled out, that man could be improved by stockbreeding methods, and that in this way a corrupt and mentally deficient aristocracy and royalty could be replaced by something better.

The public loved it. The book was banned by the Lord Chancellor. William Lawrence reluctantly went along with this, deciding that a lucrative career in the scientific establishment was preferable to bankruptcy on a matter of principle (he later became president of the Royal College of Surgeons, and accepted a baronetcy from Queen Victoria). His book remained an enormous illicit success, going through at least nine editions before 1848.

Charles Darwin later admitted having read it, although he seems not to have understood the evolutionary principles which it put forward. Other texts began to emerge, all of them anticipating Darwin in some respect, and all of them helping prepare a climate of opinion in which *The Origin of Species* could ultimately triumph. In 1831 an almost unknown Scottish botanist, Patrick Matthew, published a specialized book *On Naval Timber and Arboriculture*. Tucked away in an appendix at the back was an outline of how and why evolution had taken place which, as the author furiously pointed out in a letter to *Gardener's Chronicle* twenty years later, anticipated Darwin's theory in every essential.[2]

Matthew died as obscurely as he had lived, but to the end of his days he remained nettled by the world's refusal, as he saw it, to recognize his claim for priority in evolutionary theory. An extract from his thesis seems to justify his indignation.

> . . . it is only the hardier, the most robust, better suited to circumstance individuals, who are able to struggle forward to maturity, these inhabiting only the situations to which they have superior adaption and greater power of occupancy than any other kind; the weaker, less circumstance-suited, being prematurely destroyed. This principle is in constant action, it regulates the color, the figure, the capacities, and instincts; those individuals of each species, whose color and covering are best suited to concealment or protection from enemies, or defence from vicissitude and inclemencies of climate, whose figure is best accommodated to health, strength, defence and support . . .[3]

Whether or not Darwin read this work is not clear. Although Matthew's book is nowadays very rare, it was reviewed at the time in several journals of natural history, and is the kind of source material that Darwin, as assiduous researcher and correspondent, would have been expected to come across. Darwin, irritated and provoked by Matthew, gave very little ground to him, and wrote that this work "remained unnoticed" until his letter to *Gardener's Chronicle*.

Mr. X

However, it is certain that Darwin listened to, and noted with care, the words of someone else who saw the force of natural selection: Edward Blyth, a south London chemist a year younger than Darwin, whose passion for natural history had led him to neglect his business, so that he had to sail for Calcutta at the age of thirty-one and take up a poorly paid position as curator of vertebrate collections in the local museum.

Before leaving for India, Blyth spoke often at scientific meetings in London, some of them attended by Darwin. Darwin's early notebooks on "transmutation" (for years he avoided using the word evolution) contain transcriptions of what Blyth said. In 1835 and 1837 Blyth's theories were published in the *British Magazine of Natural History*. Darwin, according to a cryptic reference in a letter, seems to have read these too.[4]

His handling of Blyth's work is, I think, a vivid early example of how Darwin achieved success as a natural historian. On the one hand there is his familiar reluctance to identify his sources and influences, so that his masterwork seemed at the end to be entirely original. But in the context of the times, this is perhaps forgivable. It was not the Victorian fashion to acknowledge predecessors and give references in the way it is mostly done in science today. More importantly, he showed real insight by listening to Blyth and realizing that everything he was saying could be used to support an opposite conclusion.

Like creationists now (and he may well have been right), Blyth saw natural selection as a force for stability. Darwin, however, perceived that a completely different interpretation could be placed upon Blyth's observations. In the following passage, the similarity with Darwin's own writings in *The Origin of Species* is obvious; indeed, the late Loren Eiseley, in *Darwin and the Mysterious Mr. X,* has chronicled a large number of passages that are almost word-for-word identical between Darwin and Blyth. Darwin, by turning Blyth's arguments upside down, clearly showed the mark of an inquiring, agile and perceptive mind, regardless of whether we now feel he was right or wrong.

> Among animals which procure their food by means of their agility, strength, or delicacy of sense, the one best organized must always obtain the greatest quantity; and must, therefore, become physically the strongest and be thus enabled, by routing its opponents, to transmit its superior qualities to a greater number of offspring. The same law therefore, which was intended by Providence to keep up the typical qualities of a species can be easily converted by man, into a means of raising different varieties.[5]

Vestiges of Creation

Finally among Darwin's influences was one of the nineteenth century's great opinion-formers and popular educationalists, the publisher Robert Chambers. Whatever the effect of Matthew and Blyth on Darwin personally, to the general public they were unknown. Not so Chambers, whose output of dictionaries, gazetteers, atlases, educational courses, journals of knowledge and, above all, his *Encyclopaedia*, all issuing from his Edinburgh publishing house, made him a household name.

However, it was not under the Chambers imprint that *Vestiges of the Natural History of Creation* appeared in 1844. Instead Chambers, who had written the book himself, posted it surreptitiously to Manchester and thence to London, where it appeared anonymously. The reason for Chamber's caution was immediately apparent, for the book was attacked in violent terms by every scientist given the opportunity. Mostly, they took exception to the many casual errors which Chambers had made, for instance in his cavalier treatment of what distinguished one species from another. Even for its times it was an unscientific book, and scientists felt justifiable irritation.

But their chorus of disapproval, which went on for several years, aroused the opposite effect among the general public. The more the book was berated, the more they bought it. Rumors as to the identity of its anonymous author mounted—one of the most popular choices being Prince Albert, the German husband of Queen Victoria, who was known to take an interest in scientific matters. Four editions appeared within seven months, and by 1860, when *The Origin of Species* had recently reached the bookstalls, some 24,000 copies had been sold.

Chambers, with his flair for knowing what the literate working-class public (and their superiors) wanted to read, had combined two strands of thinking that achieved an immediate intuitive response. Earth, as Lyell had shown, had evolved gradually over millions of years; so, said Chambers, had life on Earth: "The simplest and most primitive type gave birth to the type next above it, this again produced the next higher, and so on to the very highest, the stages of advance being in all cases very small—namely, from one species only to another; so that the phenomenon has always been of a simple and modest character."[6]

Darwin's own copy of *Vestiges of Creation* was well thumbed, and had many marginal notes in his near-illegible hand. But whatever inspiration he drew from the text was as nothing to the service which Chambers had done him in exciting public interest in the subject of evolution; and also, like a lightning conductor, in attracting the bolts of hostile criticism upon himself.

The Rev. Adam Sedgwick, Darwin's old master at Cambridge University, where he was professor of Geology, was later to criticize *The Origin of Species*, but it was muted stuff compared with the spleen that he vented on *Vestiges*:

> . . . things must be kept in their proper places if they are to work together for any good. If our glorious maidens and matrons may not soil their fingers with the dirty knife of the anatomist, neither may they poison the springs of joyous thought and modest feeling, by listening to the seductions of this author; who comes before them with the serpent coils of a false philosophy, and asks them again to stretch out their hands and pluck forbidden fruit; who tells them that their Bible is a fable when it teaches them that they were made in the image of God; that they are the children of apes and the breeders of monsters . . . [7]

Even Thomas Huxley, although he regretted it later, was so infuriated by the unscientific inaccuracies of *Vestiges* that he was driven to such words as "foolish fancies," "charlatanerie," "pretentious nonsense."

Hiding beneath such intemperate criticism must have been a secret fear that Chambers was dangerously close to getting it right. Certainly, his public thought so. But thus far, none of the heretical authors had quite put it all together. Where they were popular, they were muddled or naïve or unscientific. Where they were coherent, as was the case with Patrick Matthew, they were obscure. Nineteenth-century biology still needed a Darwin. But thanks to his predecessors, the ground was dug and prepared. The climate, as Charles Darwin was about to launch his masterwork, was one of expectation and doubt.

Judged in his youth, Charles Darwin seemed an unlikely candidate for posterity to honor. At school, his teachers thought him below average in intellect. His father (a doctor, like grandfather Erasmus) once told him he cared for "nothing but shooting, dogs, and rat-catching, and you will be a disgrace to yourself and to your family."

Shooting, ironically for someone who was later to have had such a feeling for natural history, was his passion. The opening of the season for grouse-shooting was, he wrote to a friend, "bliss on earth." In his autobiography he wrote that, at school and afterward, "I do not believe that anyone could have shown more zeal for the most holy cause than I did for shooting birds. I used to place my shooting boots open by my bedside when I went to bed, so as not to lose half a minute in putting them on in the morning."[8]

Sent to Edinburgh University in October 1825 to study medicine, he found the lectures "intolerably" dull, and the practical work in hospital wards (particularly the sight of blood) sickening. In two years there, he completely forgot the Greek and Latin he had learned at school—or so he said. Moving to Cambridge University, he managed, just, to pick up a bachelor of arts degree to prepare him for life as a clergyman (very much a last-resort career for an upper-middle-class Englishman), now that he had rejected doctoring. Mathematics were "repugnant" to him, his attendance at lectures "almost nominal." When finally he passed his examinations he was relieved and delighted: "I'm through, I'm through, I'm through," he exclaimed to a fellow student.

However, even at this apparently unpromising stage, there were a number of indications that young Charles was marked out for better things. He had, it is clear, enormous charm. "At breakfast, wine or supper parties," wrote a contemporary, "he was ever one of the most cheerful, the most popular, and the most welcome."[9] His teachers were wrong in their estimate of his intellect, for if somewhat indolent usually, he was both painstaking and bright when interested in the subject matter, such as anything to do with the countryside.

Darwin's Gifts

Above all, there was something about his personality that made him one of those rare people capable, as his father later put it, of creating confidence in others. It was a gift that helped him throughout his life, when scientists and academics of greater scholarship freely shared their knowledge with him, and came to his defense when he was attacked. Already, at university, the characteristic was evident. He became the protégé of one of Cambridge's most esteemed dons, the Rev. J. S. Henslow,

Professor of Botany, to whose house he was often invited for family dinner, and who accompanied him on walks and boating trips in pursuit of botanical knowledge. Darwin himself, who as he grew older developed a perhaps understandable sense of self-importance, later wrote that this friendship was due to there being "something in me a little superior to the common run of youths."[10]

As to the origins of *Origin*, they can already be seen in Darwin's burgeoning fascination with all aspects of the natural world. Alongside the horse riding and shooting which delighted him so came a genuine curiosity about the countryside—its rocks, fossils, vegetation, fauna. At university, beetles fascinated him especially, and he was credited with the capture of several rare species. The summer of 1831, after he graduated, he spent on field trips near his home in Shropshire, and in the Welsh hills with another of his influential Cambridge friends, Adam Sedgwick, Professor of Geology.

On 29 August, returning home, he found a letter from Henslow saying that a Captain Fitzroy wanted to have a companion aboard H.M.S. *Beagle* for a journey to the South Seas lasting at least two years and maybe longer. Darwin, while still professing that he wished to be a clergyman, quickly saw what opportunities there would be, and made for the job headlong. Sidestepping his father's objections with the aid of his uncle Josiah Wedgwood, he then had to exert all his considerable personal charm on Fitzroy.

It was a sticky first interview. Fitzroy was autocratic, dogmatic about right and wrong, able (appointed command of the *Beagle* when only twenty-three), a Tory where Darwin was a Whig, descended illegitimately from Charles II, and a deeply convinced Christian who believed in the literal truth of every word of the Bible. Among his many mistaken convictions was the notion that a man's character could be definitively judged by the features of his face. Darwin nearly failed on this count alone because his nose, Fitzroy thought, failed to indicate energy and determination, the two qualities most needed on a long and trying voyage.

Nevertheless, Darwin won him over, and was offered the job, which was basically to keep Fitzroy happy: literally to stop the man going mad with isolation in the confines of a small ship where social and naval custom forbade the captain even to eat with anyone else on board. Darwin's presence was needed "more as a companion than as a collector," he was told. Primarily, what was called for was someone of the right social background to accompany a captain worried about being hereditarily inclined to insanity (his uncle Viscount Castlereagh, British Foreign Secretary during the Napoleonic wars, had committed suicide by cutting his throat).

On Board H.M.S. *Beagle*

So Darwin's enthusiasms and amateur skills as a naturalist were not the main reason for having him on the voyage. Although subsequently in life he always referred to himself as the ship's official naturalist, initially this was not the case. The *Beagle* already had the man for this task—the ship's surgeon, Robert McKormick, who before and afterwards performed this duty capably on other expeditions by the British Navy, notably the voyage to the Antarctic between 1839 and 1843.

However, it did not take long for McKormick to depart from the scene, and Darwin to take his place. Having started the voyage in a spirit of apparent willing cooperation, Darwin made clear his private feelings with such comments as: "My friend the doctor is an ass." On board, he let it be known that wherever the *Beagle* landed, he intended to make his own botanical and zoological collections—and in fact did so at the first two ports of call, Cape Verde and Bahia. By April 1832, only four months after they had set sail from Plymouth, McKormick was redundant. In Rio de Janeiro, he decided to go home, and Captain Fitzroy dismissed him under the guise of having him invalided out. Being invalided out, Darwin wrote home, was a euphemism for "being disagreeable to the Captain . . . he is no loss."[11]

Immense scholarship has been devoted to establishing what contribution Darwin's five years on the *Beagle* made towards the theory that he was to expound in *The Origin of Species*—what seminal ideas were formed that would later coalesce into the belief that natural selection was the overriding force in evolution. The picture that emerges is not a simple one. There was certainly no moment when he cried Eureka! and the truth was suddenly revealed. Indeed, there is a growing consensus of opinion among historians that Darwin returned from the voyage almost as scientifically bewildered as when he embarked upon it.

From South America in 1834 he wrote home: "I have not one clear idea about cleavage, stratification, lines of upheaval. I have no books which tell me much, and what they do I cannot apply to what I see. In consequence, I draw my own conclusions, and most gloriously ridiculous ones they are."

In Australia two years later, the mysteries of creation were still bothering him: "I had been lying on a sunny bank and was reflecting on the strange character of the animals in this country as compared to the rest of the World. An unbeliever in everything beyond his own reason might exclaim, Surely two distinct Creators must have been at work."[12] In between, in the Galapagos Islands where his famous finches show micro-evolution at work (they have become distinct species as a result of geographical isolation), he did not even recognize them as finches—nor, in many cases, note down which variety came from which island.

Illness

What the years on the *Beagle* undoubtedly effected, however, was a profound psychological change. They forced Darwin in upon himself, changing him from a gregarious enthusiast into a brooding invalid. The sports-loving, card-playing young graduate who left Britain's shores in December 1831 was in no way fitted for the sustained intellectual concentration that led to *The Origin of Species*, the withdrawn, often morose twenty-seven-year-old who returned in October 1836 assuredly was.

Two factors brought this change about—twin aspects of the voyage that made it, for Darwin, a prolonged saga of mental and physical misery, broken intermittently by field trips ashore. The first was Fitzroy himself. Before departure, Darwin imagined him to be "everything that is delightful." However, he soon realized how prematurely wrong this judgment was.

Fitzroy's temper was foul. In the mornings, particularly, he would stride about the ship picking fault with some minor detail or other. Junior officers would inquire whether "much hot coffee had been served out this morning"—meaning how out of sorts the captain had been.

Beyond this, Fitzroy was certifiably manic depressive. Long dark silences would descend upon him when something or somebody had offended him. In Valparaiso, his moods became so intense that they culminated in him having a nervous breakdown, when he temporarily gave up command.

For five years, Darwin was forcibly cabined and confined with

Darwin's finches from the Galapagos Islands in an illustration from his diary. Their differently evolved beaks, each perfectly adapted for its task, eventually led him to formulate his theory of the origin of different species.

this man, subjected to lectures on Biblical and other orthodoxies delivered with the insufferable arrogance of Fitzroy's caste. Darwin had to dissemble, and to hold his tongue. "The difficulty of living on good terms with a Captain of a Man-of-War is much increased by its being almost mutinous to answer him as one would answer anyone else; and by the awe in which he is held—or was held in my time, by all on board," he recollected later.

Secondly, Darwin's health deteriorated. The first symptoms, most likely psychosomatic in origin, appeared in the two months before the *Beagle*'s departure, when she was being fitted for the journey and then storms kept her in harbor. Darwin developed a rash on his hands, and chest palpitations which he thought indicated a heart complaint. He dared not have them treated, for fear that Fitzroy would find out and refuse to let him on the ship.

On board, Darwin suffered almost unrelieved physical torment, partly from the cramped conditions, and partly from seasickness. Every inch of space on the ship was packed with equipment and provisions for the journey. Darwin's hammock was first slung above the desk which he and Fitzroy shared for writing. The space was so cramped that he had to remove a drawer to make room for his feet.

Later, when he was moved into a cabin in the stern (shared with a midshipman), the more vigorous motion of the ship caused him endless seasickness, the severity of which he had never dreamed of. ''The real misery only begins when you are so exhausted that a little exertion makes a feeling of faintness come on—I found nothing but lying in my hammock did any good.''

Although Darwin once wrote to his father, perhaps to reassure him, that he found the *Beagle* ''a very comfortable house,'' and that ''if it was not for seasickness the whole world would be sailors,'' a different story emerges from his diary. He wrote despairingly of loneliness, and how much he missed the company of old friends.

> Other losses, although not at first felt, after a period tell heavily. These are the want of room, of seclusion, of rest, the jading feelings of constant hurry, the privation of small luxuries, the comforts of civilization, and last even of music and the other pleasures of imagination . . . Even the greater number of sailors, as it appears to me, have little real liking for the sea itself. It must be borne in mind how large a proportion of the time during a long voyage is spent on the water, as compared to the days in harbour. And what are the boasted glories of the illimitable ocean? A tedious waste, a desert of water, as the Arabian calls it.[13]

Small wonder that Darwin's character, after five years of such deprivations, should be transformed. Forced through circumstances to the introverted activities of collecting, reporting, musing and writing, ''I discovered, though unconsciously and insensibly that the pleasure of observing and reasoning was a much higher one than that of skill and sport.''

Two years after his return to England, this private and almost permanently unwell man, whose heart began palpitating at the least hint of mental strain, became engaged to Emma Wedgwood, daughter of the same uncle who had helped him board the *Beagle*. Nine days before their marriage, he wrote his bride-to-be a plaintive letter that reveals how aware he was of the change that had come over him. His contentment, he said, now resided entirely in leading a solitary life, and he hoped she would soon teach him that "there is greater happiness than building theories and accumulating facts in silence and solitude."

Neither to Emma nor to anyone else at this stage did he give any hint that he was on the point of achieving the conceptual breakthrough that biology needed. Perhaps he did not realize it himself.

After briefly visiting his family, Darwin's first task back in England was to arrange the geological specimens he had collected and to prepare his *Journal* of the voyage for publication. He spent the winter in Cambridge to do this, and moved to lodgings in Great Marlborough Street, in central London, in the spring of 1837. Here, as a bachelor, he spent the most creative two years of his life.

At the start, although puzzled by the enormous abundance and rich variety of different kinds of plants and animals he had seen on his travels, he was still of the orthodox Christian opinion that each of them had been created specially by God. In the course of completing his *Journal*, however, the awkward questions begged by this belief came more and more to occupy his mind. Why did some species seem to blend almost indistinguishably into one another? Why had some species been created with apparently useless organs? Why had some species become extinct, and others lived?

Origins of Origin

In July, having by now listened to Edward Blyth and rejected his conclusions, his doubts accumulated to the point where he opened a notebook on the "Transmutation of Species." He had taken the first and biggest step: he recognized that over long periods of time, species had changed and evolved. Although he later wrote, in a famous passage, that he "worked on true Baconian principles, and without any theory collected facts on a wholesale scale," this does not seem quite so. He had already adopted the theory—then an unproved one—shared by the other evolutionists we met earlier: that living things were not the same now as in the past. Next he set about gathering evidence in favor of this.

Having established in his own mind the *fact* of evolution, Darwin set to work to establish *how* it had happened. In his autobiography, he wrote that the solution came to him in a flash one evening some sixteen months later, in the autumn of 1838, while casually reading the renowned *Essay of Population* by the British economist and clergyman Thomas Malthus. Malthus argued that all the seeds of life scattered with such profusion could not possibly survive, and were kept in check only by such means as famine, disease and war. Darwin, by his own account, immediately applied the theory to nature at large:

> Being well prepared to appreciate the struggle for existence which everywhere goes on from long continued observation of the habits of animals and plants, it at once struck me that under these circumstances favourable variations would tend to be preserved and unfavourable variations would be destroyed. The result of this would be the formation of new species.

Darwin's autobiography, alas, is full of humbug. Written late in life, and intended as a moral tale for his grandchildren rather than for public consumption, it greatly oversimplifies the tortuous mental processes through which *The Origin of Species* was finally brought to the written page. Some of its statements (e.g., "I never happened to come across a single naturalist who seemed to doubt about the permanence of species") are so unreliable that even so strong a Darwinian supporter as George Simpson felt bound to comment that "they cannot literally be true, yet Darwin cannot consciously be lying, and he therefore may be judged unconsciously misleading, naïve, forgetful, or all three."[14]

Analysis of Darwin's private notebooks for this supposedly visionary month of October 1838 has shown no dramatic day of breakthrough. Instead of ! or !! or even !!! which he was accustomed to use when he felt he had come across a startling piece of evidence, there is a commonplace summary of Malthus's theory, which it is clear he was already familiar with. The next day he dropped the subject, and wrote about the sexual curiosity of apes.[15]

The habits of caution, of keeping private thoughts unspoken, which Darwin learned aboard the *Beagle*, had penetrated deep. *Origin* was a prodigiously long time in gestation. Although the bones of his theory were fully formed by the time he was married in February 1839, it was twenty years before he presented it to the public. For the time being, instead of thrusting forward with the task of collecting material which would support his idea, he occupied himself with papers for the Geological

Society (of which he became secretary); with a book on *Coral Reefs*; and with a five-volume *Zoology* classifying the specimens that had been collected on the voyage.

In 1842 he took off during the summer to write a short (thirty-five pages) penciled summary of the theory of natural selection; abandoned it for eighteen months to work on *Geology of the Volcanic Islands*; and returned to the subject again to write a treatment 189 pages long, completed in July 1844.

By now he had enough faith in it to have the manuscript copied and bound. In a letter to his wife, attached to the text, he instructed her to put aside a sum of £400 to £500 to have it edited and published, should he die before he had time to complete it. "If, as I believe, my theory in time be accepted even by one competent judge, it will be a considerable step in science."

Other Preoccupations

But once again, instead of pursuing the subject, Darwin took a step sideways. After completing a second edition of his *Journal*, and finishing *Geology*, he began in the autumn of 1846 a four-volume classification of minuscule barnacles, about the size of a pinhead, known as cirripedia. It was to occupy him, with almost obsessive irrelevance, for eight years. When he finished, he was forty-five years old, and confessed to being as sick of barnacles as any homesick sailor who felt his ship being slowed by them.

The mystery of why he took so long to face up to the job of writing *Origin* has been the subject of frequent scholarly speculation; nobody knows the answer. Evidently, he was not satisfied with his 1844 sketch, or he would not have left such a substantial sum for an editor to work on it. Also, throughout the period of gestation, he was corresponding with a motley assortment of friends, amateur naturalists, breeders and scientists, asking them for useful information that, presumably, might help settle his mind on the matter.

To close friends such as Charles Lyell or Joseph Hooker, the botanist who urged him to publish or be preempted by someone else, he was quizzical. "Do not flatter yourself that I shall not yet live to finish the Barnacles, and then make a fool of myself on the subject of species," he wrote to Hooker. Perhaps, as some have suggested, he was still so bound up in the conventions of his time that he was reluctant to put forward an anti-religious bombshell. Perhaps, on the other hand, his continuing ill-health and the enforced loneliness that he had suffered on board the *Beagle* led to a corrosive self-doubt, so that he wondered whether after all his theory was right. Perhaps, seeing the popular success of Chambers's *Vestiges of Creation*, he was biding his time, waiting for the most opportune moment.

Darwin and Wallace

At any event, prompted by Charles Lyell, Darwin began writing his masterwork in May 1856. The crisis that finally forced him into the open came in a letter on 18 June 1858, postmarked from the Malayan archipelago. Darwin, by now regarded as one of the foremost scientists in Britain, holder of the coveted Royal Society medal for biology (ironically, for the work on cirripedia that added so little to the sum of human knowledge), found in the letter a paper written by an unknown twenty-one-year-old naturalist, Alfred Russel Wallace.

It was entitled *On the Tendency of Varieties to Depart Indefinitely from the Original Type*, and Darwin read it with a dismay that rapidly turned into despair. Almost phrase for phrase it put forward a theory identical to Darwin's 1844 sketch. Already in a low mental state because of the recent death of one of his daughters, Darwin was shattered. He could not ignore the letter, for Wallace specifically asked him to pass the paper on to Lyell if he thought it "sufficiently novel and interesting"—which evidently it was, since it was virtually a carbon copy of Darwin's own thoughts on the matter. Darwin wrote that day to Lyell in a state of near panic:

> Your words have come true with a vengeance—that I should be forestalled. You said this, when I explained to you very briefly my views of "Natural Selection" depending on the struggle for existence. I never saw a more striking coincidence; if Wallace had my MS sketch written out in 1844, he could not have made a better short abstract! Even his terms now stand as heads of my chapters.

Darwin realized at once that he had no option but to offer Wallace's paper for publication, but was in despair at the thought of losing his priority. He begged Lyell for advice on whether he also could publish, now that Wallace had preempted him.

> Can I do so honourably, because Wallace has sent me an outline of his doctrine? I would far rather burn my whole book, than that he or any other man should think that I had behaved in a paltry spirit. Do you not think his having sent me this sketch ties my hands? . . . This letter is miserably written, and I write it now that I may for a time banish the whole subject; and I am worn out with musing . . . My good dear friend, forgive me. This is a trumpery letter, influenced by trumpery feelings.

As things turned out, Darwin need not have worried. Lyell and Hooker solved the immediate problem with uncommon speed. They arranged that Wallace's and Darwin's each be read, together with documents demonstrating the earlier claim of Darwin to the theory, at the next meeting of the Linnaean Society on 1 July, less than a fortnight away.

A Historical Nonevent

The meeting was something of a nonevent, historical though it became in retrospect.[16] Lyell and Hooker attended, to stress the importance of the occasion (Darwin himself, predictably, was ill), but to no effect. On a soporific summer's day, the twenty-five scholars present may have been too inattentive to take the message in. Five other papers were also read, in addition to an hour-long eulogy to a former president, recently deceased.

Certainly, the cautious and convoluted prose style of Darwin (and to a lesser extent of Wallace) disguised what the message was. It is notable that neither man dared spell out in so many words that what they were really talking about was outrageously unorthodox: evolutionary species change. Instead, they gently suggested that gradual adaptations might produce, say, a dog with more fur, or a pigeon with longer legs—''a progression to which there appears no reason to assign any definite limits.''

They left it up to the audience to draw the implicit conclusion, which was presumably that, without "definite limits," a dog may learn to fly, or a pigeon to run. And this conclusion the audience signally failed to draw. Indeed the president of the Linnaean Society wrote in his annual report at the end of the year, in a passage for which he has since been mocked, that the past twelve months had been unmarked by any revolutionary scientific discoveries.

However, the papers were now on record, and printed in the the *Journal* of the Linnaean Society (Zoological), volume 3, pp. 45–62. Darwin pressed ahead with *Origin*, working continuously for a year to expand his 1844 sketch into publishable form. It was, he stressed constantly in letters to Lyell and Hooker, only an abstract of a much larger book on the whole subject, but even so he was tormented with the difficulty of writing.

"It is an accursed evil to a man to become so absorbed in any subject as I am in mine," he wrote to Lyell; and to a cousin: "I am weary of my work. It is a very odd thing that I have no sensation that I overwork my brain; but facts compel me to conclude that my brain was never formed for thinking."

His determination to forge ahead despite the rough going, spurred on by the knowledge that competition was in the wind, had a certain ruthlessness about it. Wallace, far away on the other side of the world, was ignored. Finally, on 24 November 1859, the book was published, all 1,250 copies of the first edition having been pre-sold to booksellers. Wallace has a brief mention in the introduction. So does Darwin's friend Joseph Hooker. None of his evolutionist contemporaries or predecessors is referred to, apart from a sly unfavorable comment on the anonymous author of *Vestiges of Creation*. Darwin had become extremely protective of his brainchild. He used the words "my theory" no less than forty-five times.

Along with the Bible, *Principia Mathematica*, and *Das Kapital*, Darwin's *Origin of Species* must rank among the least read (at least, in full) and most influential books of all time. It is no exaggeration to talk of it bringing about a Darwinian revolution.

Within a decade of its publication, Darwin was able to write satisfyingly of the "now almost universal belief" in evolution. Sir Charles Lyell, after an initial hesitancy, had been won over. Darwin had been awarded the highest scientific honor England could offer, the Copley Medal of the Royal Society. Even church periodicals had modified their initial opposition, and were recommending Darwinism to their readers.

"Extinguished theologians lie about the cradle of every science as the strangled snakes beside that of Hercules," Thomas Huxley exulted.[17]

Yet by the standards of the day, *Origin* was not a huge popular publishing success. In 1872, when its sixth and final edition (in Darwin's lifetime) came out, its total sales had been 16,000, less than a third of *Vestiges of Creation*. Its fame came largely by word of mouth, passed round the clubs and high tables of the educated English public; and its appeal lay in the brilliant simplicity of its thesis, presented in a manner so weighty as to make criticism seem presumptuous.

Thomas Huxley having exclaimed how stupid he was not to have thought of the idea himself, summed up what he saw as the contribution of *Origin:*

> The facts of variability, of the struggle for existence, of adaptation to conditions were notorious enough; but none of us had suspected that the road to the heart of the species lay through them, until Darwin and Wallace dispelled the darkness.[18]

The philosopher Marjorie Grene agrees:

> What the genius of Darwin achieved, surely, was not to discover a host of new facts unknown to his predecessors that somehow added up to the further fact of evolution through natural selection; what he did was to see the facts in a new context—an imaginative context, the context of an idea, but an idea which seemed and seems to many modern minds peculiarly factual, an idea so convincing, so congenial, so satisfying that it feels like fact . . . It was the idea of natural selection that convinced the Victorians that evolution happened: so much so that for many people the idea of evolution *means* natural selection still.[19]

A Bewildering Masterpiece

Whereas nowadays all scientists would make a distinction between the fact of evolution and the theory of how it took place, Darwin deliberately connected and mingled the two. Without natural selection, he declared, evolution could not have happened; by proving natural selection, you also proved evolution. The full title of his book spelled out the message unambiguously: *The Origin of Species by Means of Natural Selection*. (In deference to his publishers, who thought this obscure, he added a subtitle: *The Preservation of Favoured Races in the Struggle for Life*.)

This confusion of two objectives makes the book extraordinarily hard to read. Darwin called it "one long argument," and apologized that "there appears to be an innate defect in my mind leading me always in the first instance to express any idea in the most awkward possible form." Yet in spite of the convoluted prose style where the theory is concerned, the book is crammed from beginning to end with reassuring and highly readable botanical and biological titbits, and Darwin implies that in the background there are thousands more such facts that he would include were the book not merely an abstract, all of them supporting the idea of natural selection as *the* creative evolutionary force.

The book thus becomes an evidential masterpiece, both bewildering and beguiling in the massive weight of its scholarship. It is quite unnecessary to read it from cover to cover to be persuaded of its central truth: that living things can inherit variations, and that because of this they have perfectly adapted to their environment. As Professor W. R. Thompson wrote in his 1956 introduction to the Everyman edition of *Origin*, "because of the extreme simplicity of the Darwinian explanation, the reader may be completely ignorant of biological processes yet he feels that he really understands and in a sense dominates the machinery by which the marvellous living forms have been produced."

A Technique for Solving Problems

Darwin also has a beguiling way of making you think he has faced up to all the objections to his theory, and overcome them. An entire chapter is devoted to "difficulties" in which he confronts such problems as the origin of flight.

In this section he begins by showing persuasively that there is a finely graded series of living squirrels, ranging from those that can only jump short distances to a variety that can glide through the air "an astonishing distance from tree to tree." Then he examines the flying lemur, where, although there are no transitional forms between them and earthbound lemurs, he "can see no difficulty in supposing that such links formerly existed, and that each had been formed by the same steps as in the case of the less perfectly gliding squirrels."

Next (having conceded that lemurs and bats aren't the same

kind of creature at all), he says there is no insuperable difficulty, by means of natural selection, in lengthening the membrane-connecting fingers and forearm of the lemur and converting it into a bat's wing. Finally, after a few digressions, he affirms that "seeing that we have flying birds and mammals, flying insects of the most diversified types, and formerly had flying reptiles, is it conceivable that flying-fish which now glide far through the air, slightly rising and turning by the aid of their fluttering fins, might have been modified into perfectly winged animals." (They weren't, of course, but that is by the by.)

Darwin's technique throughout, according to the philosopher Dr. Gertrude Himmelfarb, is to convert possibilities into probabilities, and liabilities into assets. In this particular chapter

> the solution of each difficulty in turn came more easily to Darwin as he triumphed over—not simply disposed of—the preceding one. The reader was put under a constantly mounting obligation; if he accepted one explanation he was committed to accept the next. Having first agreed to the theory in cases where only some of the transitional states were missing, the reader was expected to acquiesce in those cases where most of the stages were missing, and finally in those where there was no evidence of stages at all. Thus by the time the problem of the eye was under consideration, Darwin was insisting that anyone who had come with him so far could not rightly hesitate to go further . . .
>
> As possibilities were promoted into probability, and probability into certainty, so ignorance itself was raised to a position only once removed from certain knowledge. When imagination exhausted itself and Darwin could devise no hypothesis to explain away a difficulty, he resorted to the blanket assurance that we were too ignorant of the ways of nature to know why one event occurred rather than another, and hence ignorant of the explanation that would reconcile the facts to his theory.[20]

The Victorian reader was thereby multiply reassured. Selection encompassed everything. It showed how dogs and pigeons and orchids and sheep could be changed in character; in *Origin* there were hundreds of examples of this happening, and if more examples were needed, Darwin had them ready and waiting for inclusion in the projected expansion of his "abstract"; evolution was therefore proved; he had faced up to the difficulties posed by his theory, and shown them to be comfortingly groundless; and doubtless the fossil record, as time went on, would reveal the necessary transitional links, now that geologists knew what to look for.

But we know now that geologists haven't found the fossil intermediates, and that natural selection on its own seems unable to explain the origin of species. Darwin, probably, got it wrong. So can he still be revered as one of the giants in the history of science? Will he stay on his pedestal in the British Museum of Natural History?

There are many reasons for answering yes, and yes. First there is the sheer weight of his scholarship. From his time at Cambridge onwards, he was a supreme natural historian, observing, noting, comparing, wondering. "What a superb observer he was! And with what enthusiasm!" Ernst Mayr wrote of his *Journal of Researches*. "Not only a superb observer, he never stopped asking the searching question: Why is this or that occurring? He would propose some hypothesis or theory that might account for the phenomenon or process. And then he would make more observations that would either confirm or refute his hypothesis."[21]

Besides *Origin*, *The Descent of Man*, and a dozen other books, he wrote more than 150 papers for scientific and popular journals of his time. "These smaller essays and notes are as amazing as Darwin's books, displaying a breadth of interest that ranges from geology to all aspects of botany and zoology," Mayr commented. "Again they reveal Darwin's enthusiasm and insatiable curiosity."[22]

A Lasting Impact

It was this dogged and painstaking work that brought him recognition during his lifetime from his scientific peers. And in turn this hard-won reputation made his heretical suggestion acceptable to the general public when *Origin* came to be published. Both groups were ready for something of the kind, their appetite stimulated by others. But Darwin was the man who rose to the occasion and satisfied them.

In a sense, the weaknesses of his book were also its strengths. Wordy, repetitive, even confused and tautologous, it was still the publication that he had always said would be necessary to convert the world from a previous way of looking at life. A short paper such as his and Wallace's to the Linnaean Society went unnoticed; treatises such as Chambers's *Vestiges of Creation* were beneath consideration by anyone with scientific training.

Origin was long enough, and dense enough, to make the breakthrough. However much he may have been a man favored by the circumstances of his time, Darwin deserves credit for being the catalyst at the moment when science made one of its quantum jumps, and the public understood. It is this achievement which has come to be honored.

Sir Julian Huxley, Thomas Huxley's grandson, once said: "Darwinism removed the whole idea of God as the creator of organisms from the sphere of rational discussion."[23] Arthur Koestler wrote that despite its inadequacies, "the theory contained a basic truth: the fossil record testified that evolution was a fact, that Darwin was right and Wilberforce was wrong, so Darwinism became something of a credo for all enlightened, progressive people, while the details of the theory could be left to the experts."[24]

Above all, *Origin* provided a framework within which biology, for more than a century, has been able to work. It has been called the supreme integrative principle, and by this it is generally meant that Darwin established that life was not created in its present condition but is constantly changing—evolving. He believed also that all living things were of common descent: all mammals had one original forefather; so did all birds, reptiles, fishes, and so on. We all, in other words, have a single common ancestor and the diversity of life has come about because of branching from this source. New species emerge and survive according to the prevailing conditions on Earth.

In the sweep of this grand concept, quibbles about how species actually originated have for a century seemed somehow not to matter. Gertrude Himmelfarb wrote that

natural selection may have succeeded by default, simply because no other explanation has been available. Science, it is well known, abhors gaps as it abhors leaps, and for the same reason. The uniformity of nature and the continuum of scientific theory are both threatened by them; science's mode of knowing, its very existence, is put in jeopardy. Scientists cannot long—and a century of research is a long time as the history of modern science goes—live with the unknown, particularly when the unknown resides at the heart of their subject.[25]

Or as the naturalist F. Wood Jones wrote as long ago as 1943:

Only a fool could deny the revolutionary impact of Darwinism on the outlook of the nineteenth century, when—as one biologist put it—the educated public was faced with the alternative "for Darwin or against evolution." But the narrow sectarianism of the neo-Darwinists of our own age is an altogether different matter; and in the not-too-distant future biologists may well wonder what kind of benightedness it was that held their elders in thrall.[26]

Changing Perspectives

Thomas Huxley warned Darwin that new truths begin as heresies and end as superstitions, and that *Origin* might come to be accepted with as little justification as it had once been rejected. In 1959, at the Chicago centenary celebrating the publication of *Origin*, his words rang true in an orgy of uncritical admiration.

It was the director of the British Museum of Natural History at the time, Sir Gavin de Beer, who set the tone of the Chicago celebrations. If a layman sought to "impugn" Darwin's conclusions, he said, it must be the result of "ignorance or effrontery."[27] Sir Julian Huxley, having said that natural selection could explain everything, added: "We do not intend to get bogged down in semantics and definition."[28] Garrett Hardin of the California Institute of Technology said that anyone who failed to honor Darwin "inevitably attracts the psychiatric eye to himself."[29]

As 1982 approached, and with it the centenary of Darwin's death, the atmosphere was altogether more restrained. In 1980–81 two major scientific occasions showed just how far away from the theory of natural selection biology was moving. The first was a four-day conference, again in Chicago, called simply "Macroevolution." The other was a new display, "Origin of Species," in the British Museum of Natural History, to mark its own centenary.

Sir Gavin de Beer had died and his place at the museum was taken by Ronald Hedley, who began to reshape the exhibitions there. "Origin of Species" seems at first glance an innocuous display—nothing so immediately upsetting as cladism and the abandonment of the search for man's ancestors. Instead, it presents a bland run-through of Darwinian theory from Darwin's time to our own, superficially leaving the impression that everything is solved and natural selection reigns triumphant still.

But when you read the small print, you find that doubt runs deep. Note the "may be," the "almost instantly," and the "many of them involve natural selection" (i.e., not all) in the following passage from the handbook accompanying the exhibition:

> We have seen that the evolution of new species may be a slow process involving natural selection. There is evidence amongst living species for the different stages of this process. We have also seen how new species can arise almost instantly by a change in the chromosomes, called polyploidy. Polyploidy is particularly important in the formation of new plant species. There are also other theories about how new species are formed, and many of them involve natural selection at some stage. But ideas about evolution are continually changing as new evidence is added . . .

It is a far cry from the gradualism which Darwin insisted upon, and which is crucial to the neo-Darwinist synthesis. In the concluding passage, the retreat from the view of say, Sir Gavin de Beer and Sir Julian Huxley that natural selection, and natural selection alone, is responsible for evolutionary change, is made even more explicit:

> Today it is generally accepted that other mechanisms, some not yet fully understood, may also have played their part in the evolution of new species. Since Darwin's time, evolutionary theory has been expanded and modified, with new evidence continually being added from molecular biology, population dynamics and, in particular, from genetics. The theory of natural selection, and the debate surrounding it, have stimulated an enormous amount of research and raised a great many questions. What factors influence the changing patterns of variation amongst living things?
>
> What role can fossils play in helping us to interpret evolutionary change?
>
> Does chance play an important part in evolution?
>
> These are just a few of the questions still to be answered.

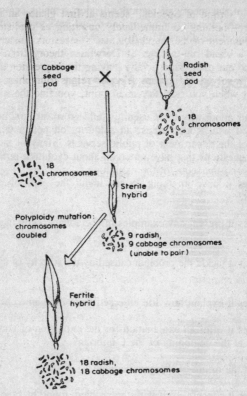

Cabbage
seed
pod

X

Radish
seed
pod

18
chromosomes

18
chromosomes

Sterile
hybrid

Polyploidy mutation:
chromosomes
doubled

9 radish,
9 cabbage chromosomes
(unable to pair)

Fertile
hybrid

18 radish,
18 cabbage chromosomes

A polyploid organism has acquired, through mutation, at least twice the basic number of chromosomes. Polyploidy in plants is fairly common, and gives rise to the sudden appearance of dramatically new and different species, which then multiply by self-reproduction (as in the example of the fertile hybrid radish/cabbage example in the diagram). Many scientists think that some kind of similar chromosomal rearrangement may be the source of species change among mammals (as in Goldschmidt's "hopeful monsters" in chapter six). The problem is that polyploid organisms cannot reproduce with anything except their own kind. A polyploid animal would therefore be sterile unless there was a population of identical polyploids with which to mate.

But the theory of natural selection remains central to any study of evolution and is one of the keys to our understanding of the diversity of life.

Dialogue With a Museum

Curious about why the museum would go so far, and no further, I talked with Roger Miles, an authority on pollen who is also head of the department of public services. Many of the troublesome questions that have emerged about evolution turned out to be beyond the scope of the exhibition.

Does it have anything to say about the gaps on the fossil record?

No.

Does it present any examples of a gradually evolving series of fossils?

No.

Does it tackle the problem caused by the scarcity of transitional forms?

No.

Does it explain how life emerged from inorganic chemicals?

No.

Does it offer an explanation for the explosion of complex life forms at the beginning of the Cambrian?

No.

Nor the origin of the genetic code?

No.

Does it concern itself with the problem found by breeders, that there is a genetic limit to change?

Breeding is mentioned in the way that Darwin saw it, as an analogy to evolution.

What does it say about the origin of flight?

Nothing.

Does it touch common patterns of form, such as segmentation?

No.

What about Goldschmidt's hopeful monsters—embryonic restructuring?

Nothing.

Anything about punctuated equilibria?

No.

Extinctions?

Nothing.

Catastrophes?

Nothing again.

Is natural selection supposed to work fastest when competition is intense, or relaxed?

We don't get into it.

Lamarckism?

Only that he was one of Darwin's predecessors.

And so on. The British Museum of Natural History, evidently, is hardly the hotbed of revolutionary change envisioned by its critics in the correspondence and editorial columns of *Nature*.

Roger Miles, a mild-mannered man, pointed out reasonably that the exhibition was aimed at an interested lay public who would know nothing about Darwinism but had probably heard of Darwin himself. "If you are aiming at people who are starting absolutely from scratch, you have to get to first base before you start querying natural selection too much. For working evolutionary theorists we may seem over-dogmatic on the establishment side. I hope not. We'll have to see.

"You have to remember that while we must remain open to other possibilities, we also have to tell people things in a coherent way. The way we do this is to accept the standard line at a particular moment but hedge it about to a certain extent, saying this doesn't seem to be the only system, or there is room for doubt here, and people are doing research there. But you still have to tell a story.

"We've tried to be honest—to say that there are other ways of looking at it, that the establishment view may not be the only one. We've certainly come a long way from our last exhibition on evolution, the one de Beer put together twenty years ago. He wrote a handbook in which it was said that these days, evolution is accepted as a fact, and natural selection is the mechanism for it, full stop. As far as he was concerned, the interesting conceptual bit of it was completely wrapped up, there was nothing left to think about.

"Now, through all the holes that are being picked in Darwin's theory, I think we are coming towards a better understanding. It's a matter of proportions. At the moment, everyone seems to be proposing a particular theory as if it's the only answer. But they needn't be mutually exclusive. Darwin and Lamarck can live together."

The Gradualism Dispute

On the other side of the Atlantic, things were less gentlemanly. Gathered together at the Field Museum of Natural History for the conference on "Macro-evolution" were almost all the renowned names we have come across in this book, together with an assortment of other geologists, palaeontologists, molecular biologists, embryologists, and—the old guard—population geneticists.

"Clashes of personality and academic sniping created palpable tension in an atmosphere that was fraught with genuine intellectual ferment," reported the journal *Science*. "The proceedings were at times unruly and even acrimonious."[30]

The central issue was, once more, gradualism. The palaeontologists were unanimous in insisting that the fossil record does not consist of the gradual change which Darwin hoped for, and neo-Darwinism needs. Instead, it shows stasis and extinction.

Stephen Gould said so forcibly. "For millions of years species remain unchanged in the fossil record. Then they abruptly disappear, to be replaced by something that is substantially different but clearly related." The geneticist Gabriel Dover from Cambridge called species stasis—their tendency to persist over a long period unchanged—"the single most important feature of macro-evolution."

Reaction from the population geneticists was mixed. Some refused to accept the evidence. Leonard Stebbins, a veteran synthesist, continued to insist that "you don't have to invoke anything except the natural selection of small differences." But Francisco Ayala, who ranks with George Simpson, Ernst Mayr and Theodosius Dobzhansky in the United States as a teaching evolutionary theorist of the traditional school, seemed converted: "We would not have predicted stasis from population genetics, but I am now convinced from what the palaeontologists say that small changes do not accumulate."

If gradualism goes, and the pattern is revealed as one of sudden leaps, a new mechanism for change is needed. Speakers floated many of the ideas described earlier in this book: hopeful monsters, embryological change, chromosomal jumps, laws of biological form. "It is easier to identify the issues than to draw conclusions." David Raup summed up. But *Science* was less enigmatic: "The central question of the Chicago conference was whether the mechanisms underlying micro-evolution can be extrapolated to explain the phenomena of macro-evolution. At the risk of doing violence to the positions of some of the people at the meeting, the answer can be given as a clear No."

Those who see in this turmoil the death throes of Darwinism may be underestimating the monster's capacity for survival. It has long been a criticism of Darwin's great theory that by seeming to explain so much, it actually explains very little. Because of the tautology that lies at its heart, it can be made to encompass limitless subsidiary theories without damage to its massive structure.

Darwin himself, at the end of his life, adopted a form of Lamarckism. Since then, Darwinism has swallowed within itself a succession of detailed discoveries about how life is passed on from one generation to the next: Mendelism, the nature of chromosomes, the structure and function of genes. On the one hand you can say, like *Nature*, that this is striking evidence of its overwhelming consistency, and that no theory of such a grand scope in the physical sciences has done as well in the past century.[31] On the other hand you can complain, like C. D. Darlington, that Darwinism began as a theory that evolution could be explained by natural selection, and ended as a theory that evolution could be explained as you would like it to be explained.[32]

There are signs that this process of absorption is taking place once more. The modern synthesists, John Maynard Smith foremost among them, are pointing out that ideas concerning the development of form, and differing rates of change, were put forward in their writings twenty-five years ago and more.[33] This may be true, if somewhat less than candid. The emphasis throughout that time has been overwhelmingly ·on the technical and mathematical austerities of chance and necessity.

Darwin, I feel sure, would have felt happier with the new approach; for in a sense it is not a new one at all. Unlike those who pore over the barren statistics of population genetics, Darwin and the great naturalists of Victorian times never lost sight of the creative wonder of nature as a whole. The marvelous closing passage of *Origin* shows him at his finest.

He contemplates ''an entangled bank, clothed with many plants of many kinds, with birds singing on the bushes, with various insects flitting about, and with worms crawling through the damp earth.'' Then he reflects how "these elaborately constructed forms, so different from each other, and dependent on each other in so complex a manner, have all been produced by laws acting around us.''

And he concludes: "There is grandeur in this view of life, with its several powers, having been originally breathed into a few forms or into one; and that, whilst this planet has gone cycling on according to the fixed law of gravity, from so simple a beginning endless forms most beautiful and most wonderful have been, and are being, evolved.''

The new biology is looking afresh at living things—at their shapes, their patterns, their dynamics and their relationships. If, after more than a century, natural selection has been tested and found wanting, and if we are left once again with a sense of ignorance about origins, Darwin would not have minded. Science is a voyage of discovery, and beyond each horizon there is another.

References

Chapter One

1 David M. Raup, "Conflicts between Darwin and Paleontology," *Bulletin*, Field Museum of Natural History, 50, Jan. 1979.
2 ibid.
3 See Gish, 1978, and Anderson/Coffin, 1977.
4 Nilsson, 1954.
5 Attenborough, 1979, p. 112.
6 Berril, 1955, p. 10.
7 Ommaney, 1964, p. 60.
8 Romer, 1966, p. 36.
9 Norman D. Newell, "The nature of the fossil record," *Proc. Amer. Phil. Soc.*, 103, 2 (1959), pp. 264–85.
10 Daniel I. Axelrod, "Early Cambrian marine fauna," *Science*, 128 (1958), pp. 7–9.
11 T. N. George, *Science Progress*, 48 (1960), p. 1.
12 G. L. Stebbins, "Evolution as the central theme of biology," *Biological Sciences Curriculum Study Newsletter*, 49 (Nov. 1972), p. 4.
13 D. V. Ager, personal communication.
14 *New Scientist*, 5 April 1979, p. 25.
15 e.g. in Simpson, 1949.
16 Butler, 1911.
17 David B. Kitts, *Evolution* 28 (1974), p. 467.
18 F. H. T. Rhodes, "The course of evolution," *Proc. Geol. Ass.*, 77, 1, 1966.

Chapter Two

1 Mayr, 1963, p. 586.
2 In Waddington, 1969, pp. 82–9.
3 "Is a new and general theory of evolution emerging?," *Palaeobiology*, 6, 1 (1980), pp. 119–30.
4 Leakey, 1979, p. 30.
5 Quoted in Macbeth, 1974, p. 36.

6 Koestler, 1974, p. 113.
7 Huxley, 1953, p. 50.
8 Wilson *et al.*, 1973, pp. 793–4.
9 *American Scientist*, 45 (1957), p. 385.
10 Dobzhansky, 1962, p. 140.

Chapter Three

1 Ayala and Dobzhansky, 1964, p. 307.
2 ibid.
3 Hoyle and Wickramasinghe, 1978.
4 Harold F. Blum, *American Scientist*, 43 (1955), p. 595.
5 Fox, 1965, p. 359.
6 M. J. E. Golay, "Reflections of a communications engineer," *Analytical Chemistry*, 33 (June 1961), p. 23.
7 F. B. Salisbury, *American Biology Teacher*, 33 (1971), p. 335.
8 J. Little, *New Scientist*, 4 September, 1980.
9 Ayala and Dobzhansky, 1964, pp. 116–17.
10 Williams, 1966, quoted in Ho and Saunders (see chapter six, ref. 12).
11 Dawkins, 1976, p. 12.
12 Barash, 1979, p. 21.
13 J. B. S. Haldane, "Population genetics," *New Biology*, 18, pp. 34–51.
14 Quoted in Ho and Saunders (see chapter six, ref. 12).
15 G. Webster and B. Goodwin, *History and Structure in Biology*; apply to authors at University of Sussex.
16 M.-W. Ho and P. T. Saunders, "Beyond neo-Darwinism—an epigenetic approach to evolution," *J. Theor. Biol.*, 78 (1979), pp. 573–91.
17 Grassé, 1977.
18 Personal communication. Experiment reported in S. E. Luria and M. Delbruck, "Mutations of bacteria from virus sensitivity to virus resistance," *Genetics*, 18 (1943), pp. 491–511.
19 T. Maniatis and M. Ptashne, "A DNA operator-repressor system," *Scientific American*. June 1976, pp. 64–76.
20 Duncan and Weston-Smith, 1977, pp. 210–17.
21 Medawar and Medawar, 1977, p. 39.
22 E. Mayr, "The evolution of Darwin's thought," *Guardian*, 22 July 1980, p. 18.
23 Chain, 1970.
24 Moorhead and Kaplan, 1967.
25 ibid.

26 ibid.
27 R. Lewontin, "Adaptation," *Scientific American*, September 1978, pp. 119–30.
28 J. S. Morton, *Impact 90*, Institute for Creation Research, San Diego.
29 See ref. 20.
30 C. P. Martin, "A non-geneticist looks at evolution," *American Scientist*, January 1953, p. 101.
31 Dobzhansky, 1951, p. 73.
32 Patterson, 1978, p. 65.
33 See ref. 22.
34 Patterson, 1978, p. 70.

Chapter Four

1 e.g., Wysong, 1976, p. 308.
2 Patterson, 1978, p. 142.
3 Attenborough, 1979, p. 243.
4 Darwin, 1859.
5 de Beer, 1964.
6 *Enc. Britt.*, "Evolution," p. 619.
7 Heberer and Wendt, 1977, pp. 345ff.
8 e.g., Simpson, 1949, p. 119.
9 Desmond, 1977, pp. 180–209.
10 Gould, 1978.
11 ibid.
12 Quoted in T. Bethell, "Darwin's mistake," *Harper's Magazine*, February 1976.
13 Based on Himmelfarb, 1959.
14 Mayr, 1963, p. 12.
15 Simpson, 1964, p. 81, quoted in Macbeth 1974, p. 25.
16 Moorhead and Kaplan, 1967, p. 14.
17 Goldschmidt, 1940, summarized in his "Evolution as viewed by one geneticist," *American Scientist*, 40 (1952), pp. 84–123.
18 Personal communication.
19 Pamphlet no. 142.
20 Simpson, 1949, pp. 168ff.
21 See ref. 19.
22 Gould, 1978.

Chapter Five

1 *Nature*, 239 (October 1972), p. 420.
2 *Nature*, 240 (December 1972), p. 365.

3 "Scientific Creationism," presented 16 November 1978.
4 *Impact,* no. 65, Institute for Creation Research.
5 *Creation News Sheet,* no. 18, 3 Church Terrace, Penylan, Cardiff.
6 Whitcomb and Morris, 1961.
7 ibid, pp. 232–3.
8 See ref. 3.
9 Nelkin, 1977, p. 89.
10 Segraves, 1975, p. 30.
11 *Impact,* no. 81, Institute for Creation Research.
12 Personal communication.
13 *Nature,* 285 (May 1980), p. 130.
14 Segraves, 1975, pp. 14–15.

Chapter Six

1 G. L. Bush, "Modes of animal speciation," *Annu. Rev. Ecol.,* 6 (1975), pp. 339–64.
2 Goldschmidt, 1940, p. 390.
3 ibid., pp. 390–91.
4 Sussex University, December 1980.
5 "The return of hopeful monsters," *Natural History,* 86 (1977), p. 30.
6 ibid.
7 Interviewed on BBC TV *Horizon,* March 1981.
8 G. N. Bush *et al.,* "Rapid speciation and chromosomal evolution in mammals," *Proc. Nat. Acad. Sci.,* 74 (1977), pp. 3942–46.
9 White 1978, p. 8.
10 Quoted in Cannon, 1959.
11 Quoted in Steele, 1979, p. 8.
12 M.-W. Ho and P. T. Saunders, "Adaptation and natural selection: mechanism and teleology," paper presented at the conference on "The Dialectic of Biology and Society in the Production of Mind," Bressanone, Padua University, March 1980.
13 Steele, 1979, p. 36.
14 Personal communication.
15 *Trends in Biochemical Sciences,* 5, 3, p. xv.
16 *Nature,* 290 (1981), pp. 508 and 513.
17 Ager 1980, pp. 42ff.
18 "Fossils in evolutionary perspective," *Science Progress,* 48, 189 (January 1960), p. 2.
19 Velikovsky, 1972, p. 77.

20 *Catastrophist Geology*, 1, 1 (June 1976), p. 5.
21 Hoyle, 1979.
22 *Scientific American*, 208 (1963), p. 77.
23 Schopf, 1972, p. 84.
24 See ref. 7.
25 See chapter one, ref. 1.
26 "The chance that shapes our ends," *New Scientist*, 5 February 1981, p. 349.
27 *Proc. Geol. Ass.*, 89 (1977), p. 100.
28 Ager, 1973, p. 100.
29 "A theory of evolution above the species level," *Proc. Nat. Acad. Sci. USA*, 72, 2 (February 1975), p. 650.
30 *New Scientist*, 23 April 1981, 7 May 1981, 21 May 1981.
31 Personal communication.
32 *New Scientist*, 4 October 1979, p. 40.
33 Maynard Smith, 1966.
34 See ref. 12.
35 *Journal of Physics A.*, Math. & Gen., 11, no. 10 October 1978, pp. 2107–30.
36 *SIS Review*, IV, 1, autumn 1979, p. 8. Apply Bernard Trescott, 12 Dorset Road, Merton Park, London SW19.
37 Hapgood, 1970.
38 Sanderson, 1969, pp. 103–16.
39 *SIS Review*, IV, 2/3, winter 1979–80, p. 65.
40 Foley, 1976, p. 115.
41 Nilsson, 1953, p. 1198.
42 Moore, 1940, p. 143.
43 Ager, 1973, p. 41.
44 Velikovsky, 1956, pp. 190–91.
45 Hitching, 1978, p. 13.
46 ibid.
47 ibid.
48 *Nature*, 229 (1971), pp. 553–4.
49 *Guardian*, 11 September 1980, p. 11.
51 ibid.
51 *The Times*, 10 July 1980, p. 18.

Chapter Seven

1 Personal communication.
2 Quoted in chapter three, ref. 15; see also Driesch, 1914.
3 Simpson, Pittendrigh and Tiffany, 1957, p. 472, quoted in Koestler, 1975, p. 145.
4 Koestler, 1975, p. 148.

5 Jantsch and Waddington, 1976, p. 13.

6 Waddington, 1976, p. 13.

6 Waddington, 1957.

7 See Saunders, 1980.

8 Personal communication.

9 Personal communication.

10 Personal communication.

11 "In the game of energy and thermodynamics you can't even break even," *Smithsonian Institute Journal*, June 1970, pp. 6ff., quoted in Morris 1975.

12 ibid.

13 Morris, 1975, p. 123.

14 M.-W. Ho, Open University lecture on evolution, September 1980.

15 See ref. 5, p. 44.

16 See ref. 14.

17 ibid.

18 *Physics Today*, 25, 2 (1972).

19 Quoted by E. Jantsch, personal communication.

20 Lars Lofgren, in his paper "Knowledge of evolution, evolution of knowledge."

21 Elise Boulding, in her paper "Evolutionary visions, sociology and the human life span."

22 Personal communication.

23 Personal communication.

24 Personal communication.

25 Personal communication.

26 *J. Theor. Biol.*, 86 (1980), pp. 757–70.

27 Personal communication.

Chapter Eight

1 Richard Wrangham, "The Bishop of Oxford: not so soapy," *New Scientist*, 9 August 1979, p. 450. Also BBC-TV, *Monitor*, November 1980.

2 Quoted in Assmuth, 1914, p. 63.

3 ibid., p. 10.

4 ibid., pp. 14–15.

5 Eiseley, 1959, p. 272.

6 Quoted in Bowden, 1977, pp. 46–7.

7 ibid., p. 46.

   "Morton's ranking of races by cranial capacity," *Science*,
      '341 (May 1978), p. 504.

9 "Whatever happened to Zinjanthropus?" *New Scientist*, 26 March 1981, p. 802.

10 "Cladistics," *Biologist*, 27, 5 (1980), p. 234.

11 British Museum Natural History, 1980, p. 5.

12 ibid., p. 20.

13 *Nature*, 288 (20 November 1980), p. 208.

14 *Nature*, 288 (4 December 1980), p. 430.

15 See ref. 13.

16 *Nature*, 289 (26 February 1981), p. 735.

17 See also P. Forey, "Introduction to Cladistics," apply to author at British Museum of Natural History.

18 N. E. Platnick, "Philosophy and the transformation of cladistics," *Systematic Zoology*, 23, pp. 446–51.

19 Patterson, 1978, p. 151.

20 See ref. 10, p. 239.

21 Gould, 1978.

22 "The secrets of yesterday's man," *Guardian*, 23 April 1981, p. 11.

23 Quoted in Davidheiser, 1969, p. 240.

24 Waddington, 1969, p. 10.

25 "Evolution by orderly law," *Science*, 164 (May 1969), p. 684.

26 Bowden, 1977, pp. 124–48.

27 ibid.

Chapter Nine

1 "The origin of Darwinism," *Scientific American*, July 1964; see also Darlington 1950.

2 *Gardener's Chronicle*, 7 April 1860.

3 ibid.

4 Eiseley, 1979.

5 ibid.

6 see ref. 1.

7 ibid.

8 Barlow, 1958.

9 Darwin, 1887.

10 See ref. 8.

11 Gould, 1978, "Ladders, bushes and human evolution."

12 ibid.

13 Darwin, 1887.

14 "Charles Darwin in search of himself," *Scientific American* August 1959.

15 S. J. Gould, "Darwin's deceptive memories," *New Scientist*, 21 February 1980, pp. 577–9.

16 J. W. T. Moody, "The reading of the Darwin and Wallace papers—a historical non-event," *J. Soc. Bilphy. Nat. Hist.*, 5, 6, pp. 474–6.

17 Quoted in Himmelfarb, 1959.

18 ibid.

19 Grene, 1966, pp. 192–3.

20 Himmelfarb, 1959, pp. 274–5.

21 "The evolution of Darwin's thought," *Guardian*, 27 July 1980.

22 ibid.

23 Tax and Callender, 1960, vol. III, p. 45.

24 Koestler, 1978, p. 179.

25 See ref. 20.

26 Quoted in Koestler, 1978, p. 204.

27 See ref. 23.

28 ibid.

29 ibid.

30 R. Lewin, "Evolutionary theory under fire," *Science*, 210 (21 November 1980), pp. 883–7.

31 "How true is the theory of evolution?" *Nature*, 290 (12 March 1981), pp. 75–6.

32 See ref. 1.

33 See ref. 30.

Books Cited in Text

Ager, Derek V., *The Nature of the Stratigraphic Record*, Halsted Press, New York, 1981.

Anderson, J. Kerby, and Coffin, Harold G., *Fossils in Focus*, Zondervar, Michigan, 1977.

Assmuth, J., *Haeckel's Frauds and Forgeries*, London, 1918.

Attenborough, David, *Life on Earth*, Little, Brown, Boston, 1981.

Ayala, F. J., and Dobzhansky, T. (eds.), *Studies in the Philosophy of Biology*, University of California Press, Berkeley, 1974.

Barash, D., *Sociobiology: The Whisperings Within*, Souvenir Press, London, 1979.

Barlow, Nora (ed.), *The Autobiography of Charles Darwin*, W. W. Norton, New York, 1969.

Beer, Gavin de, *Atlas of Evolution*, Nelson, London, 1964.

Berril, N. J., *The Origin of Vertebrates*, Oxford University Press, 1955.

Bowden, M., *Apemen—Fact or Fallacy?* CLP Publishers, 1979.

British Museum of Natural History, *Man's Place in Evolution*, Cambridge University Press, New York, 1981.

——*The Origin of Species*, London, 1981.

Butler, Samuel, *Evolution Old and New*, Gordon Press, New York.

Clark, R. E. D., *Darwin: Before and After*, Folcroft, Folcroft, PA, 1977.

——*The Universe: Plan or Accident*, The Paternoster Press, London, 1949.

Cannon, H. G., *Lamarck and Modern Genetics*, Greenwood Press, Westport, CT, 1975 (reprint of 1959 ed.).

Chain, E., *Responsibility and the Scientist in Modern Western Society*, Council of Christians and Jews, London, 1970.

Daly, Reginald, *Earth's Most Challenging Mysteries*, Presbyterian and Reformed Publ. Co., Phillipsburg, NJ 1972.

Darlington, C. D., *Darwin's Place in History*, Basil Blackwell, Oxford, 1950.

——*The Little Universe of Man,* George Allen and Unwin, London, 1978.

Darwin Charles, *The Origin of Species,* New American Library (Mentor), New York.

Darwin, Francis (ed.), *The Life and Letters of Charles Darwin* (3 vols.), Johnson Reprint Co., New York (reprint of 1888 edition).

Davidheiser, B., *Evolution and Christian Faith,* Presbyterian and Reformed Publ. Co., Philadelphia, 1969.

Dawkins, R., *The Selfish Gene,* Oxford University Press, New York, 1976.

Dewar, D., and Shelton H. S., *Is Evolution Proved?,* Hollis and Carter, London, 1947.

Desmond, Adrian J., *The Hot-Blooded Dinosaurs,* Dial Press, New York, 1976.

Dobzhansky, Theodosius, *Genetics and the Origin of Species,* Columbia University Press, New York, 1951.

——*Mankind Evolving,* Yale University Press, New Haven, 1962.

Driesch, H., *The History and Theory of Vitalism,* Porcupine Press, Philadelphia, 1982.

Duncan, R., and Weston-Smith, M. (eds.), *The Encyclopaedia of Ignorance,* Pergamon Press, New York.

Eiseley, Loren, *Darwin's Century: Evolution and the Men Who Discovered it,* Doubleday, New York, 1958.

——*Darwin and the Mysterious Mr. X,* Harcourt Brace Jovanovich, New York, 1981.

Foley, G., *The Energy Question,* Penguin, London, 1976.

Fox, S. W. (ed.), *The Origins of Prebiological Systems and their Molecular Matrices,* Academic Press, New York, 1965.

Gish, Duane T., *The Fossils Say No!,* Creation-Life Publishers, San Diego, 1978.

Goldschmidt, Richard B., *The Material Basis of Evolution,* Yale University Press, New Haven, 1982.

Gould, Stephen Jay, *Ever Since Darwin: Reflections in Natural History,* W. W. Norton, New York, 1979.

Grassé, P. P.,*Evolution of Living Organisms,* Academic Press, New York, 1977.

Grene, M., *The Knower and the Known,* Faber and Faber, London, 1966.

Hapgood, Charles H., *The Path of the Pole,* Chilton, Philadelphia, 1970.

Heberer, G., and Wendt, H. (eds.), *Grizmek's Encyclopaedia of Evolution,* Elsevier Scientific Publishing Co, Amsterdam, 1977.

Himmelfarb, G., *Darwin and the Darwinian Revolution,* Von Nostrand Reinhold, New York, 1959.

Hitching, Francis, *The World Atlas of Mysteries,* Collins, London, 1978, and Pan, London, 1979.

Hoyle, F., and Wickramasinghe, C., *Lifecloud,* J. M. Dent, London, 1978.

——*Diseases from Space,* J. M. Dent, London, 1979.

Huxley, Julian, *Evolution in Action,* New American Library, New York, 1953.

Huxley, Thomas Henry, *Lay Sermons, Addresses and Reviews,* New York, 1871.

Jantsche, E., and Waddington, C. J. (eds.), *Evolution and Consciousness,* Addison-Wesley, Reading, Massachusetts, 1976.

Keith, Arthur, *Darwin Revalued,* Watts and Co., London, 1955.

Koestler, Arthur, *The Ghost in the Machine,* Random House, New York, 1982.

——*Janus: A Summing Up,* Random House, New York, 1979.

Leakey, Richard E., *The Illustrated Origin of Species,* Hill and Wang, New York, 1982.

Macbeth, Norman, *Darwin Re-tried,* Gambit, Ipswich, MA, 1979.

Maynard Smith, John, *The Theory of Evolution,* Penguin, New York, 1976.

Mayr, Ernst, *Animal Species and Evolution,* Belknap Press of Harvard University Press, Massachusetts, 1963.

Medawar, P. B. and J. S., *The Life Science,* Harper & Row, New York, 1978.

Moore, E. S., *Coal: Its Properties, Analysis, Classification, Geology, Extraction, Uses and Distribution,* New York, 1940.

Moorhead, P. S., and Kaplan, M. M. (eds.), *Mathematical Challenges to the neo-Darwinian Interpretation of Evolution,* Monograph no. 5, Wistar University Press, Philadelphia, 1967.

Morris, Henry M., *The Troubled Waters of Evolution,* Creation-Life Publishers, San Diego, 1975.

Nelkin, Dorothy, *Science Textbook Controversies and the Politics of Equal Time,* Massachusetts Institute of Technology, Cambridge, Mass., and London, 1977.

Nilsson, H., *Synthetische Artbildung,* Lund University, Gleerup, Sweden, 1954.

Ommaney, F. D., *The Fishes,* Time-Life Publications, New York, 1964.

Patterson, Colin, *Evolution,* Cornell University Press, New York.

Penny, Douglas A., and Waern, Regina, N., *Biology*, Pitman, London, 1965.

Romer, A. S., *Vertebrate Paleontology*, University of Chicago Press, 1966.

Sanderson, Ivan T., *More "Things,"* Pyramid, New York, 1969.

Saunders, Peter T., *An Introduction to Catastrophe Theory*, Cambridge University Press, 1980, New York.

Schopf, T. M. (ed.), *Models in Paleontology*, Freeman Cooper, San Francisco, 1972.

Segraves, Kelly, L., *The Great Dinosaur Mistake*, Beta Books, San Diego, 1975.

Simpson, George Gaylord, *The Meaning of Evolution*, Yale University Press, New Haven, 1949, revised edition 1976.

——*This View of Life*, Harcourt Brace and World, New York, 1964.

——, Pittendrigh, C. S., and Tiffany L. H., *Life: An Introduction to Biology*, New York, 1957.

Steele, E. J., *Somatic Selection and Adaptive Evolution*, State Mutual Book, New York, 1980.

Tax, S., and Callender, C. (eds.), *Evolution After Darwin* (3 vols.), University of Chicago Press, 1960.

Waddington, C. H., *The Strategy of the Genes*, George Allen and Unwin, London, 1957.

——(ed.) *Towards a Theoretical Biology 2: Sketches*, Edinburgh University Press, 1969.

White, Michael J. D., *Modes of Speciation*, W. H., Freeman, San Francisco, 1978.

Williams, G. C., *Adaptation and Natural Selection: A Critique of Some Current Evolutionary Thought*, Princeton University Press, 1966.

Velikovsky, I., *Worlds in Collison*, Pocket Books, New York, 1980.

——*Earth in Upheaval*, Pocket Books, New York, 1980.

Whitcomb, John C. Jr., and Morris, Henry M., *The Genesis Flood*, Presbyterian and Reformed Publ. Co., Nutley, New Jersey, 1961.

Wilson, E. O., *et al.*, *Life on Earth*, Sianauer Associates, Sunderland, Massachusetts, 1973.

Wysong, Randy L., *Creation-Evolution: The Controversy*, Inquiry Press, Michigan, 1976.

Introductory Reading

Chapter One

Darwin's Theory The handbook accompanying the British Museum of Natural History's exhibit "The Origin of Species" is a concise and graphic guide to the theory of natural selection. At a more advanced level see their publication by Patterson.

The Fossil Record Gish 1978 and Anderson/Coffin 1977 are well-argued accounts of the fossil gaps, with strongly creationist views. So is Dewar, Douglas, *The Transformist Illusion*, Dehoff Publications, Murfreesboro, Tennessee, 1965. Kerkut, G. A. *The Implications of Evolution*, Pergamon Press, Oxford, 1960, is a critique by a highly respected biologist.

Allopatric Speciation Mayr 1963.

Chapter Two

Mendelism and Genetics Again see Patterson. Hoagland, Mahlon B., *The Roots of Life*, Avon Books, New York, 1979, is a lucidly written layman's guide. Medawar and Medawar, 1978, is another good introduction.

Neo-Darwinism Simpson, G. G., *The Major Features of Evolution*, Columbia University Press, New York, 1953, and Dobzhansky, 1951, are classic statements of the theory written at a time when it seemed proved beyond dispute. See also Maynard Smith, J., *On Evolution*, Edinburgh University Press, 1972.

Chapter Three

Origin of Life Calder, Nigel, *The Life Game*, BBC Publications, 1973 describes most of the modern theories except Prigogine's. Watson, Lyall, *Lifetide*, Hodder and Stoughton, London, 1979, is a brilliantly ambitious attempt to unite physics, biology and parapsychology.

Sociobiology Wilson, E. O., *Sociobiology: The New Synthesis*, Harvard University Press, 1975, is the standard work; Dawkins, 1976 is more readable.

Chapter Four

Flaws in Darwinism Macbeth, 1979, is a witty mock-legal attack on many aspects of the theory, particularly the tautology. See also Grene, 1966. Wolsky, M. de I., and Wolsky, A., *The Mechanism of Evolution*, S. Karger, New York, 1976, is a profound study. Grassé, 1977, reflects long-held European doubts about Darwinism. Koestler, 1982 and 1979, are essential reading. So is Gould, 1979.

Chapter Five

Creation v. Evolution Wysong, 1976, is a comprehensive study of the arguments on both sides. Creationist literature can be obtained from Creation-Science Research Center, P.O. Box 23195, San Diego, CA 92123; Institute for Creation Research, 2100 Greenfield Drive, P.O. Box 2666, El Cajon, CA 92021; Evolution Protest Movement, 13 Argyle Avenue, Hounslow, Middlesex TW3 2LE. Anti-creationist literature is published in Biological Sciences Curriculum Study, P.O. Box 930, Boulder, Colorado 80306 (Director: William V. Mayer). Nelkin, 1977, documents the background to the debate.

Chapter Six

Hopeful Monsters Goldschmidt, 1982. For a summary paper, see his "Evolution, as viewed by one geneticist," *American Scientist*, 40, 1 (1952); and ref. 5, chapter six.

Chromosomal Speciation Bush, G. L. *et al.*, "Rapid speciation and chromosomal evolution in mammals," *Proc. Nat. Acad. Sci.*, 74, pp. 3942–6 (1977). White, M. J. D., *Modes of Speciation*, W. H. Freeman, San Francisco, 1978, surveys the many possible alternative mechanisms of evolution. Gould in ref. 3, chapter two, describes the theory briefly and gives a good list of references.

Lamarckism Steele, 1980, and Cannon, 1975. See also Cannon, H. G., *The Evolution of Living Things*, Manchester University Press, 1958.

Catastrophism Ager, 1981, and his "The nature of the fossil record," *Proc. Geol. Ass.*, 87, 2, pp. 131–59 (1976). Velikovsky's work still supported in *SIS Review*, quarterly publication of the Society for Interdisciplinary Studies, 12 Dorset Road, Merton Park, London SW19; and in *Kronos*, Glassboro State College, Glassboro, NJ 08028. See also Gallant, René, *Bombarded Earth*, John Baker, London, 1964.

Punctuated Equilibria The two key papers are in Schopf, 1972, and Gould and Eldredge, "Punctuated equilibria: the tempo and mode of evolution reconsidered," *Paleobiology* 3, pp. 115–51 (1977). The theory is discussed at length in Hallam, A. (ed.), *Patterns of Evolution as Illustrated by the Fossil Record*, Elsevier Science Publishing Company, Inc., New York, 1977.

Chapter Seven

Rational Morphology Little recent scientific work exists. For Webster and Goodwin's study apply to authors at Sussex University. Koestler, 1982 and 1979, are again essential. Besides Driesch, 1914, see Thompson, D'Arcy W., *On Growth and Form*, Cambridge University Press, New York, 1917 and 1942.

Self-organizing Universe A highly original approach to the origin of form through so far undiscovered "morphogenetic fields" is put forward in a lucid book by Sheldrake, Rupert, *A New Science of Life*, J. P. Tarcher, Los Angeles, 1981.

Epigenetics Waddington's work is best summarized in his *The Evolution of an Evolutionist*, Cornell University Press, 1975.

Catastrophe Theory Saunders, 1980.

Prigogine and the Evolutionary Vision Jantsch, E., *The Self-Organizing Universe*. Pergamon Press, New York, 1980; also his edited proceedings of the symposium in Los Angeles, published by AAAS, 1981.

Chapter Eight

Fossil Apemen Reader, J., *Missing Links*, Little, Brown & Co., Boston, MA, 1981. is a non-partisan account of current rivalries. Bowden, 1977, is the best creationist book on the subject.

Cladistics Patterson's paper in *Biologist* (ref. 10, chapter eight) is a succinct introduction: the most comprehensive survey is by Peter Forey, who can be contacted at the British Museum of Natural History.

Chapter Nine

Darwin's Contemporaries Eiseley, 1959; also Gillispie, C. C., *Genesis and Geology*, Harvard University Press, 1951.

Darwin and The Origin of Species Keith, 1955, is a good biographical introduction. Himmelfarb, 1959, is essential; much criticized at the time for suggesting that Darwin's contribution had been overrated.

INDEX

Abbeville, 102–103
acid molecules, 33
acquired
  characteristics, 122–25,
    143
Acts and Facts, 91
adenine, 32
Africa, x, 22, 142, 185–86
Agassiz, Louis, 128, 129
Ager, Derek, 129–32, 135,
    137, 143–45
agriculture, 100–101
Albert, Prince, 200–01
algae, 49–50
Alvarez, Luis, 142
allopatric speciation, 24, 29
altruism, 51, 53, 57
American Anthropological
    Association, 92–93
American Association for the
    Advancement of Science,
    142, 158, 162–64
American Biology Teacher,
    49–50, 100
American Indians, 113–14
American Museum of Natural
    History, 135, 136, 181–82,
    191–92
American Scientific Affiliation,
    93–94
amino acids, 45–50 *passim*
  proteins and, 52–54
ammonia, 45–46, 48

amoeba, 29
amphibians, 7, 139, 143
  fish and, 10–11, 20
Analytical Chemistry, 47, 49
Andrews, Peter, 169, 191–93
animal husbandry, 100, 101
Antarctic, 45
anteaters, 149, 150
Antennarius hispidus see
  walking fish
anthracite, 138
Anthropogeny
  (Haeckel), 173–74
antibiotics, 36–37; 81–2
apemen, 107ff
  fossils, 96
  invented, 177–80
apes, man and, 176, 186,
    187
Archaeoceti, 69–70
Archaeopteryx, 20–23
  as bird, 96
  as intermediate, 21–23
Ark, 110–12
  Adam and, 106, 112
Asimov, Isaac, 158–60
Atlantic Ocean, 142
Atlas of Evolution (de Beer),
    16, 27, 79–80, 223
Attenborough, David, 12–13
Australia, 49, 50, 135, 149,
    150–51, 204–05
  aboriginals, 113–14, 184–85

australopithecines, 187–88,
    192
Australopithecus afarensis,
    191–92
axolotl, 117–18
Ayala, Francisco, 224

bacteria, 6–7, 14–15, 36–38,
    45, 46–47, 50–51,
    56–57, 61, 124
mutations in, 56, 81–82
Balanovski, Eduardo, 157–63
Baldwin effect, 130
baleen, 70–71
barnacles *see* cirripedia
barn owls, 73
bats, 74–75, 77
    flight, 74–77, 80–81
    Darwin on, 215–16
    pre-adaption and, 82–83
Bavaria, 21
Beagle, HMS, 194–96, 204–10
    passim
bears, 77–80
beech trees, 137
de Beer, Sir Gavin, 16, 27,
    78–80, 83–84, 173–74,
    219–20, 220, 223
Atlas of Evolution, 16, 26,
    78–80, 223
beets, sugar content
    raised, 38–39
beetles, 203–04
Berkeley, UCL, 68–69, 95,
    116, 118–19, 162–63
Berlin, 175, 178
Berril, N.J., 13–14
Bible, 113–14, 213–14
    Adam, Ark and, 110–14
    creationism and, 91–102,
        103–4, 105–6, 129–131
    Genesis Flood and, 101,
        102–4, 128–29

Velikovsky and, 131–32
binomial theorem, 57
birds, 7
    Archaeopteryx and, 21–23,
        96
    flight, 74–75, 76, 80–81
    fossils, 74–75, 112–13
    mammals and, 187–88
    reptiles and, 80–81, 187–88
    wing of, 76, 116
Birmingham
    University, 30
Biston betularia (peppered
    moth), 36–38
    changes in wing colour, 5–6,
        36–37, 42–43
bivalves, 30
    Jurassic, 30
blue whale, 45
Blyth, Edward, 198–200,
    208–9
boas, 71, 74
boatzin, 22
Bock, Walter, 175
bombardier beetle (*Brachinus*),
    67–68
Brachinus (bombardier beetle),
    67–68
brain growth, 117–19
Brigham Young University,
    22
British Association for the
    Advancement of Science,
    3–4, 171
*British Magazine of Natural
    History*, 198–99
British Museum of Natural
    History, 16, 63, 113,
    114, 152, 169, 216–17
    cladism and, 185–91, 193
    Evolution handbook, 28,
        29, 77, 191, 223

British Museum of Natural
History *(cont.)*
  'Man's Place in Evolution'
    exhibition, 184–91, 193,
    219–20
  'Origin of Species' display,
    219–223
Britten, Roy J., 58, 57, 61
Brno, 32
Broom, Dr. R., 72–73
Buckland, William, 129
Bryan, William Jennings, 93, 182
Burbank, Luther, 38
Bush, Guy, 116–17, 119

Calaveras, 102–3
calcium, 8
Calhoun, John, 163–64
California, 52, 94–95, 105–6,
    112
Californian Board of Education,
    91, 92–93, 112–13
California Institute of
    Technology, 57, 61,
    219–20
California, University of, 142
Cambrian, 8–10, 13–14, 15–16,
    45–46, 135, 222
  fossils, 14–19, 95
Cambridge University,
    50–51, 56, 69–70,
    200–201, 202–3,
    216–17, 224
Canada, 149–50
Carboniferous, 74–75
  coal measures, 137–38
Castenedolo, 102–3
Castlereagh, Viscount, 203–4
catastrophism, 128–32, 137–39,
    passim, 139, 141–42,
    143, 146–47, 194–95
  Cuvier and, 128–29, 132,
    194–95

Velikovsky and, 131–32,
    134, 138
caterpillars, 116
cats, 149–50
centipede, 75, 76, 77
Chain, Professor Sir Ernst, 64
Chambers, Robert, 199–201,
    211, 217–18
Chesterton, G.K., 180
Chicago
  centenary of *Origin of
    Species*, 219–20
  macro-evolution conference
    (1980), 30, 219–20
chimpanzees, 57, 60
  man and, 118–20
  skull of, 119
chitin, 8
Choukoutien, 180, 181
Christian Heritage College,
    94–95
Christianity Today, 91–93
chromosomes, 40, 58, 143,
    224–25
  jumps of, 118–19
  polyploidy, 221–20
  speciation, theory of,
    118–20, 139–40
Cincinnati, University of, 89
cirripedia, Darwin's
    classification of, 210–11
cladism, 29, 185–93,
    passim cladogram, 187
clams, 14–15, 22
coal
  formations, 101–02, 132
  origin of, 137–38
coelacanth, 20, 23–24,
    187–88
comparative anatomy, 6–7,
    166
Colorado, 22
Columbia University, 43–44

continental drift, 6–7,
128–29, 142, 148–49
cornea, 66–67
coordinated variables, 67–68
Copernicus, ix, 84–86
coral, 12–13, 138
Coral Reefs (Darwin), 209
corti (organ of hearing), 72–73
creationism, 91–114, *passim*,
115, 156
Adam and, 112
Ark and, 110–11
evolution and, 91–114,
137, 166
teaching of, 91–92, 93–95,
96, 104–5, 106, 112,
113–14
Creation Research Society, 93
Creation-Science Research
Center, 94, 106
Cretaceous, 135, 140, 141,
148–50
Crick, Francis, 56
crocodiles, 80–81
Cro-Magnon, 185–86, 192
crossbreeding wheat, 38–39,
121
crossopterygian fish, 11, 19–20
crustacea, 14–15, 70–71
cuttle fish, 79
Cuvier, Baron, 121,
128–29, 132, 146–48,
195
cytosine, 32–33
Czechoslovakia, 32

daman (*Hyrax*), 19–20
Darlington, Professor C.D.,
197–98, 224–25
Darrow, Clarence, 92–93,
105–6
*Darwin and the Mysterious Mr
X* (Eiseley), 198–99

Darwin, Charles, ix–x,
et passim, 194–226
Beagle voyages, 194–95,
196, 203–4, 205–8, 209,
210, 211
Blyth and, 198–200, 208–9
contemporary influences,
205–10
education, 201–4
finches, 204–5
health, 205–6, 207–8
Lyell and, 194–96, 210
marriage, 207–8, 210
predecessors' influence,
194–97, 223–24
Royal Society, 211,
213–14
Wallace and, 211–14
works
Cirrepedia, classification
of, 210–11
Coral Reef, 209
Descent of Man, The,
172–73, 216–17
*Geology of the Volcanic
Islands,* 210
Journal, 207–8 *passim,*
216–17
Origin of Species, The,
passim, 209–10, 212–26
'Transmutation of Species',
208–9
Zoology, 209
Darwin, Emma, 207–8, 210
Darwin, Erasmus, 195–97,
201–2
Darwinism, ix–x *et passim*
throughout
Das Kapital, 213–14.
dauermodifications, 127–28
Dawkins, Richard, 51
DDT, 37–38
Delbruck, 56

Descent of Man, The (Darwin), 172–73, 217

Devonian, 135

Devonian Catskill delta, 138

Dewar, Douglas, 72–73, 93–94

dinosaurs, 3, 6–7, 22–23, 70–71, 77–80, 81–82, 111, 139, 149–50, 186–87, 188, 190
   creationism and, 102–3, 106, 112
   extinction of, 140–42, 164
   flight and, 74–75, 76, 80–82, 214
   fossils, 21, 74–75
   mammals and, 140
   man and, 95

divine creation doctrine, ix, 171–72
   see creationists

DNA, 32–33, 35, 41, 45–46, 47, 49, 51–53, 55, 57, 61, 122–23
   cells and, 57, 61
   mutations of, 56, 58
   protein and, 56
   structure of, 56

Dogma of Evolution, The (More), 89

Dobzhansky, Theodosius, 43–44, 45–46, 59, 84, 224

dogs, domestication of, 39

Driesch, Hans, 148, 156

Drosophila melanogaster (fruit fly), 5–6, 39, 55
   mutations, 5–6, 39, 40, 41, 43–44, 154

Dubois, Eugene, 177–79, 181–82, 183, 185, 192

duck-billed platypus, 104

ear, mammalian, 70–73, 147–48

natural selection and, 73
   section through, *71*
   stereophonic hearing and, 73

Early Cambrian fossils, 15

earthworm, 75, 77

Eddington, Sir Arthur, 52

Eden, Murray, 64–65

Edinburgh University, 86, 202

einkorn grass, 39

Einstein, Albert, ix

Eiseley, Loren, 198–99

Eldredge, Niles, 136, 143–44, 152, 164–65, 185

elephants, 3

embryology, 116–19, 127, 151–54, 165–66

Encyclopaedia Britannica, 75, 80

Encylopadia of Evolution (Grizmek), 75

Encyclopaedia (Chambers), 200

Encyclopaedia of Ignorance, 57, 58

environmental selection, 56

environmental stimulus, 130–31 *passim*

enzymes, 52–54

Eohippus, 17–20 passim

epigenetic landscape, 151

Epihippus, 17

Epperson v. the State of Arkansas, 105

Equus nevadensis, 17

Equus occidentalis, 17

erosion, 128–29, 195

Escherichia coli, 36–37

Essay on Population, (Malthus), 208–9

Ever Since Darwin (Gould), 85

*Evolution of Living
    Organisms* (Grassé), 146
Evolution Protest Movement,
    72–73, 92–93
*Evolution: Science Falsely
    So-called*, 92–94
Evolution (Patterson)
    (British Museum of Natural
    History), 28, 29, 77
evolution, theory of, ix–x
    et passim throughout
    creationism and, 91–114
Evolutionary novelties, 66–68
*Evolution: The Modern
    Synthesis*, 75
eye, 66–68, 72–73, 116
    Darwin on, 66–67, 75–77,
        78–79
    evolution of, 78–79
    mole and, 121
    Patterson on, 75–77
    pre-adaption and, 82–83
    section through, 67
    of trilobite, 95
    of whale, 70–71

Faraday, Michael, 194
Field Museum of Natural
    History (Chicago),
    10–11, 139
    macro-evolution conference,
        224–25
finches, 204–5, *206*
fins, development of, 14
Fisher, Sir Ronald, 86–87
fish, 7, 116–17, 187–88
    amphibians and, 10–11
    crossopterygian, 10–11
    evolution of, 12–14
    fossils, 10–11, 100
    pre-adaption and, 81–83
    stasis and, 30

Fishes, The (Ommaney),
    13–14
Fitzroy, Captain, 203–4,
    206–7
flight, 74–77, 78–81
    dinosaurs and, 74, 75, 76,
        80–81, 82
    origin of, 74–75, 214–15,
        222
    reptilian, 120
Flood, 128–29, 135–37,
    195–96
    creationism and, 101–3
flounder, 116–17
flying lemurs, 214
    Darwin on, 215–16
foot and mouth virus, 45
Forey, Peter, 191
fossils, 3, 15, 19, 34, 67–68,
    100, 220
    apemen, 96
    Cambrian, 14–19, 95
    dinosaurs, 74
    of early invertebrates, 11
    evolution of horse and,
        16–17
    evolutionary novelties and,
        67–68
    first, 49–50
    graveyards, 132
    insects, 74–75
    jawbones, 11–12
    of marine life, 14–15,
        23–24
    pre-adaption and, 81–83
    Pre-Cambrian, 8–10, 19
    species change, 224
    whales in, 69–70
fossil record, 4–9 *et passim*,
    42–43, 45, 82–83, 97–98,
    112, 188, 190, 224
    catastrophism and, 129–31
    chart, 5

fossil record (*Cont.*)
  creationism and, 101–2,
    112–13
  gaps in, 5–29, *passim,* 30,
    74–75, 77–80, 80–81,
    95, 96, 103, 104, 115,
    116–17, 120, 135,
    143–44, 193, 216–17
  giraffes in, 152
Fox, Professor Sidney, 47
French Academy of Sciences,
  121
Froelich, Herbert, 164
frogs, 118–19
fundamentalism, 91–94,
  104–6
fruit fly (*Drosophila
    melanogaster*), 5–6, 39, 55
  mutations, 5–6, 39, 40,
    41, 44, 154

Galapagos Islands, 204–5
Galileo, ix
gamma rays, 133
Gardener's Chronicle, 198
Genesis Flood, The (Morris and
    Whitcomb), 101–3
genes, 32–34, 42, 43, 45–46,
    53–54, 83, 87,
    127–28, 143–44, 145,
    146–47
  conservatism and, 39
  evolution and, 59
  flow, 57, 61
  function, 225
  Groczyanski/Steele paper on,
    125–29
  laws of, 166–67
  Mendel and, 32
  mutations of, 35, 52, 58–59,
    147–48, 151–54
  rate, 116–17
  recombination of, 35

sociobiology of, 51–53
structure of, 56
substitution, 41, 42–43
Weismann's barrier and,
  54–55, 122–23
*Genetical Theory of
    Natural Selection, The*
    (Fisher), 86–87
*Genetic Basis of Evolutionary
    Change, The*
    (Lewontin), 45
genetic code, 32–33, 45, 47,
  49, 50–53, 87, 185, 222
  cells and, 57–58
  evaluation v. creation, 98–99
  formation of, 53–54,
    55–56, 61
  limitations, 38–39
  mutations and, 64–65,
    118–19
  repair mechanism, 42
genetics, 32–34, 43–44, 65,
  85, 97, 120, 220
  advances in, 39
  assimilation, 130
  changes, 42, 61, 64,
    69–70, 115, 116–17, 122
  drift, 62–63
  evolution theory and, 56
  homeostasis population,
    33–34, 42–43, 57, 61, 63,
    122–23, 148–49,
    188–90, 225–26
Geological Society, 194–95,
  209
Geologists' Association, 16,
  136, 139
*Geology of the Volcanic
    Islands* (Darwin), 210
George III, 195–96
George, Neville, 131
George Washington
    University, 52

Germany, 68, 69, 116, 137–38

germ cells, 54–55, 57, 61, 122

gills, 81–82, 118, 173–74, 175
  development of, 12–13
  lungs and, 10–11

giraffes, 3, 5–6, 85
  neck length, 30, 63, 152–53

Gish, Duane T., 95–96, 100–101, 104–5

Glomar Challenger, USS, 142

Golay, Marcel J.E., 47–50

Goldschmidt, Richard B., 69, 116–17, 118–20, 154, 221, 222

Goodwin, Brian, 156, 164, 165–67

Gothenburg event, 133

Gould, Professor Stephen Jay, 12, 15, 34, 42, 69, 85–86, 117, 130, 144, 182, 185, 192
  neo-catastrophism and, 136, 139, 140–43
  pre-adaption, 81–83
  stasis and, 224

gradualism, 136, 139, 188, 190, 220, 224–25
  dispute, 224–25

Grand Canyon, 4

Grassé, Professor Pierre-Paul, 55, 146

Gray, Asa, 66–67, 78–79, 85

Great Britain, 72–73, 92–93, 103–4, 182

Great Dinosaur Mistake, The (Segraves), 106

Grene, Marjorie, 214

Gribbin, John, 141–42

Grizmek, 75–77

Groczynski, Reg, 122–23, 124–26

ground hogs, 149–50

guanine, 32–33

Gulf of St. Lawrence, 135

Haeckel, Professor Ernst, 25, 172–75
  falsifications by, 173–78, 182–83

Haldane, Professor, J.B.S., 51–53

Hallam, Professor Anthony, 30

Halstead, Beverly, 188, 190

Hamilton, Dr. Roger, 152

Hapgood, Charles, 134

Hardin, Garrett, 219–20

Harvard University, 12–13, 23–24, 29, 34, 36, 65, 136, 139

Hedley, Ronald, 220

Henslow, Rev. J., 202

heredity, 56, 57, 61, 146

Heribert-Nilsson, Professor N., 11–12, 16, 22, 137

Hesperopithecus, 181–82

Himmelfarb, Dr. Gertrude, 215–16, 218–19

Hiroshima, 59

Hitler, Adolf, 68–69, 116

Homo erectus, 179, 185, 188, 190

Homo habilis, 191–92

Homo sapiens, 139, 188, 190

homologous organs, 147–50

Hooker, Joseph, 210, 212–13, 214

'hopeful monsters', 68–69, 116–17, 118, 120, 139, 143, 154, 221, 222

Hopi Indians, 135

horse, 83–86, 87
  evolution of, 16–20, 24, 96
  toes of, 16–17
housefly, 36–37
Hoyle, Sir Fred, 46–47, 135
Hutton, James, 195–96
Huxley, Sir Julian, 42, 75,
    77, 92, 93, 217–18, 219–
    20,
Huxley, T.H., 3–4, 7–9, 42,
    84, 86, 171–74, 177, 180,
    194–95, 202, 213–14,
    217–18, 219–20
hydrogen, 30, 45–46
  peroxide, 67–68
hydroquinone, 68
Hyrax (daman), 17, 19–20

Ichthyostega (amphibian genus),
    11
Iceland, 141–42
Illustrated London News,
    181–82, 183
immunology, 122–27, 166,
    185
Impact, 52, 106
Imperial College (London), 157
*Implications of Evolution,
  The* (Kerkut), 1
India, 198–99
Industrial melanism, 36–37
industrial pollution, 36–37
  smoke control, 37–38
'Inheritance of Acquired
    Immunological Tolerance
    to Foreign Histocompati-
    bility Antigens in Mice'
    (Steele/Groczynski),
    125–26
insects, 127–28
  flight, 74–75, 77–81
  fossils, 74–75
  segmentation in, 148

Institute for Creation
    Research, 52, 91, 94–95,
    100–101, 102
Institute for Enzyme Research,
    92
Introduction to Evolution
    (Moody), 75
invertebrates, 12–13, 120
  early sea creatures, 10–11
  fossil record and, 100
  multi-celled, 6–7, 9–10
Invincible, HMS, 110
iridium, 142
iris, 78–79

Jantsch, Professor Erich,
    162–65
Java, 177, 179–81, 183
Java Man, 172–73, 177–80,
    185, 192
jawbones
  fish development of, 12–14,
    81–83
  mammalian, 11–12, 70–73
  reptilian, 11–12, 70–73,
    74–75, 148
jellyfish, 3, 5–6, 7, 15, 19,
    33, 45–46, 115, 147–48
  eye of, 78–79
Jena University, 177
Jenkin, Fleeming, 31
Johansen, Don, 183
Johns Hopkins University,
    144–45
Jones, F. Wood, 219
Journal (Darwin), 207–8
  second edition, 210
Journal of Physics, 132, 133
Journal of Researches
    (Mayr), 216–17
*Journal of Theoretical
    Biology*, 146

Jurassic, 23
  bivalves, 30

Kalmus' ebony fly, 40
Kehoe, Alice B., 92–93, 104–6
Kenya, 191–92
Kerkut, Professor G.A., 1,
Koenigswald, G.H.R. von,
    178–79
Koestler, Arthur, 151, 218

Lamarck, Jean Baptiste de
    Monet de, 115, 120–21,
    122–23, 124–27,
    152–53, 154, 196, 197,
    198, 223–24, 225
Lawrence, William, 197–98
lens, 66–67, 77, 78, 79
Lewontin, Richard, 45, 65
Life on Earth
    (Attenborough), 12–13
Linnaean Society, 212–13,
    217–18
  Journal of, 212–13
limestone, 8, 21, 23
lions, 15, 115, 148, 153
living systems, behaviour of,
    158–61
lizard, four-legged, 74–75
London, University of, 56
'Lucy', 183–84, 185–86
Lund University (Sweden),
    12, 137
lungfish, 20, 187–88
Luria, S.E., 56
Lyell, Sir Charles, 129–31,
    194–96, 200–201, 210–13,
    214
Lysenko, 121

macro-evolution, 34–36, 37–38,
    42–44, 56, 68–69, 116,
    117–19, 166, 224–25

Madagascar, 20
Mae-Wen Ho, 130, 146, 156
magnetic field, 132
Major Features of Evolution
    (Simpson), 188, 190
Malthus, Thomas, 208–09
mammals, 6–7, 13–14,
    68–69, 70–71, 77, 127–28,
    152, 185–86
  chromosomes in, 121
  dinosaurs and, 140
  ear, 70–71
  features of, 20, 23
  jaw of, 11–12
  reptiles and, 11–12, 20,
    23, 70–73
man
  ancestry of, 189
  dinosaurs and, 95
  fossil apes and, 186–7,
    192
  genetic make-up, 60
  pedigree of, 25
  skull, 117, *119*
  soles of feet, 127–28
'Man's Place in Evolution'
    exhibition, 184–91, 193,
    219–20
Man's Place in Nature
    (Huxley), 172–73
marble, 8
Marquette University
    (Milwaukee), 92–93
marsupials, 148–49
Martin, C.P., 59
Marxism, 121, 190–91
Massachusetts Institute of
    Technology, 64–65, 105
*Material Basis of Evolution,
    The* (Bush), 116–17
Mathematical
  catastrophes, 155

'Mathematical Challenges to the Neo-Darwinian Interpretation of Evolution' (Wistar symposium), 64

Matthew, Patrick, 198, 199, 201–2

Mayas, 131

Mayr, Professor Ernst, 24, 29, 34, 36, 41, 43–44, 57, 61, 62, 69–68, 78, 83–84, 86, 120, 216–17
  Ark and, 111, 224
  cladistics and, 188, 190

McGill University, 59

McKormick, Robert, 204–5

Medawar, Sir Peter, 86–87, 124, 125

Mendel, Gregor, 32, 33, 144–45, 197–98

Mendelism, 32, 33–34, 132, 224–25

mesazoic, 22

methane, 45–46, 48

Miami University, 49

mice, 121, 122–25, 149–50
  Toronto experiments, 122–29 *passim*, 125

micro-evolution, 37–38, 42–43, 56, 204–5, 224–25

Miles, Roger, 222–23

Miller, Stanley, 45–46, 48

Minnesota, University of, 95

Missing Links (Reader), 171, 173

mole, life of, 121, 149–50

molecular biology, 3–4, 55–6, 60, 126–27, 143–44, 145, 220

molluscs, 30, 138

monkeys, 32-33, 117–19

Moody, Paul Amos, 75–77

Moore, E.S., 138

More, L.T., 89

Morgan, T.H., 39, 83–84

morphology, 146–148

Morris, Henry M., 91, 95, 101–102

Morton, Dr. Jean Sloat, 52

mosquito, 118

moths *see* peppered moth

multi-celled invertebrates, 6–7, 9–10, 57
  fossils of, 10–11, 15–19

Mungo event, 133

Museum of Natural History (Paris), 120

mustard gas, 59

mutations, 33–34, 43–44, 61–64, 122
  in bacteria, 56, 81–82, 154
  chance, 61–65, 82–83, 122, 150–51, 165–66
  evolution and, 57–61
  eye and, 67–68
  fish and, 116–117
  fruit flies and, 5–6, 39–42
  of genes, 35, 43–44, 52, 59, 57–61, 147–48, 150–54
  inherited, 58
  'neutral', 62–63
  neo-Darwinism and, 122–123
  'point', 56, 118–20
  polyploidy and, 221
  random, 130, 143–44
  rate of, 57–61, 128–29
  snake and, 71–74

myxomatosis, 37–38

NASA, 48, 47–49

Natchez, 102–3

National Museum of Natural Sciences (Ottawa), 141

Natural History of Creation (Haeckel), 173–74

Natural History of Man
    (Lawrence), 197–98
natural selection, 3, 7–9 *et passim* throughout, 24–29
  considered, 85–87
  evolutional oddities and, 67–68
  mammalian ear and, 73
Nature, 46–47, 91–92, 110, 124–126, 184, 190-91, 193, 194, 223, 224–25
Nature of the Stratigraphical Record, The (Ager), 136
Neanderthals, 177, 185–86, 192
Nebraska Man, 181–83
Nelkin, Dorothy, 105–106
'Nellie' *see* Pekin Man
neo-catastrophism, 135–39
neo-Darwinism, 5, *et passim*
Neolithic man, 39
Newell, Professor Norman D. 135–139
New Scientist, 20–23, 124–26
Newton, Isaac, ix–x, 84–86, 91–92, 164–65, 194
Niessen, Richard, 106
nucleic acids, 47–49
nucleotides, 32–33, 35, 45–46
Nutcracker Man, 172–173

octopus, 14–15, 45–46
Olmo, 102–3
Ommaney, F.D., 13–14
On Naval Timber and Arboriculture (Matthew), 198
Ontario Cancer Institute, 122
On the Tendency of Varieties to Depart Indefinitely from the Original Type (Wallace), 211
Oparin, A.L., 48

Open University, 146
Origin of Species, The, ix, 3, 3–4, 7–9 *et passim* throughout
  centenary of, 219–20
  preliminary work for, 209–10
Orohippus, 17
Ordovician, 13–14
Origin of Vertebrates, The (Berril), 13–14, 31
Osborn, H.F., 181–82
ostrich, 22
  callosities of, 128–30
Ostrom, Professor John H., 22
Oxford University, 51–53
oyster, 23–24, 24–29, 30

Palaeocene, 13–14
palaeontology, 3–4, 19–20, 20–23, 29, 30, 49–50, 97, 143–44, 188–90
  Darwinism and, 11–13
  fossils and, 15–19, 120
Paramys genus, 13–14
Paris, University of, 64–65
Patterson, Colin, 63, 75–77, 113–114, 186–88, 190–92
peat, 137
Pekin Man, 172–73, 180–82, 192
penicillin, 61–64
Pennsylvania, University of, 61–64
peppered moth (*Biston betularia*), 36–38
  changes in wing colour, 5–6, 36–37, 42–43
Permian, 135, 139
pharyngeal pouches, 175
Philosophie Zoologique (Lamarck), 120
photosynthesis, 141
phyla, 15–19

Physics Today, 157, 162–63
pigeons
  Darwin and, 213
Piltdown Man, 172–73,
    181–83, 183–84
Pithecanthropus, 179
Pithecanthropus alanthus,
    177–80
*Pithecanthropus erectus, a
    Humanlike Transitional
    Form from Java* (Dubois),
    180
Pitman, Michael, 69–70
placental animals, 148–49
plants, 112–13, 127–28, 139,
    141
  domestication, 39
Platnick, Norman, 191–92
Plato, 135
Pliocene, 181–82
point mutations, 56
Pollard, Jeffrey, 56–57,
    122–23
polyploidy, 220–21, *221*
Popper, Sir Karl, 124–26, 157,
    163–64
population
  change, 33–34
  dynamics, 220
  genetics, 33–34, 42–43,
    57–61, 63, 122–23, 146,
    188–90, 225–26
pre-adaption, 81–84
Pre-Cambrian, 8–10, 13–14,
    57–61
  lack of fossils, 8–10,
    15–19
Prigogine, Ilya, 157–58,
    passim 160–164
Principia Mathematica,
    213–14
Principles of Geology
    (Lyell), 129–31, 194–96

*Proceedings of the National
    Academy of Sciences* (US),
    122–23
proteins, 47–49, 52
  clock, 185
  DNA and, 56, 58
proto-fish, 12–13
pteranodons, 74–75
pterodactyls, 74–75, *76*
pterosaurs, 74–75, 81–82
punctuated
  equilibria, 136–39
pupil of the eye, 66–68
pygostyle bone, 21
pythons, 71–74

Queen Elizabeth College, 146

rabbits, 37–38
radiometric dating, 97–100,
    101–2
ramapithecines, 187–188
rats, 37–38
Raup, David M., 10–11, 139,
    224
Reader, John, 171, 172–73,
    184
Reading University, 188–190
Reagan, President, 104–6
recapitulation, theory of,
    173–75
Rendel's yellow fly, 40
reptiles, 6–7, 80–81,
    143–44
  Archaeopteryx and, 21–23
  birds and, 80–81, *81*
  ear of, 71
  flight and, 120, 215
  fossils, 100, 112–13
  jaw of, 11–12
  mammals and, 11–12, 20–23,
    71–72, *73*
  snakes and, 71, 74

retina, 66–68, 70–72, 79
rhinoceros, 118
Rift Valley (Kenya), 152
RNA, 45, 122–23
rodents, 13–14
rods and cones, 66
Romer, A.S., 13–14
Rousseau, Jean Jacques, 120
Royal College of Surgeons, 197–8
Royal Society, 211
    Copley medal of, 213–14
Royal Zoological Society, 72–73, 194–95
Russell, Dale, 141

Saint Louis University, 91–92
salamander, 57, 118
Salisbury, Frank B., 49
Sanger, Frederick, 56
Sarich, Vincent, 185–86
Saunders, Peter, 130, 146–47
scales, 81
Schindewolf, Otto, 116–17
Schrödinger, Erwin, 160–61
Schutzenberger, Marcel P., 64–65
Science, 224, 225
Science Progress, 131
Scopes, John, 181–83
    trial of, 92–93, 105–06
sea food, 6–7
sea lilies, 14–15
sea snails, 45–46
sea urchins, 14–15
Sedgwick, Rev. Adam, 200–202, 203–4
Segraves, Kasey, 106, 112
Segraves, Kelly L., 94–95, 110–11, 112
selective breeding, 38–39
    of fruit flies, 41–42

Selfish Gene, The (Dawkins), 51
sexual reproduction, 57, 61
sharks, 79
shellfish, 23
shrimps, 79
Siberia, 121, 132, 134–35
silicon, 8–9
Simpson, Professor George Gaylord, 23–24, 78, 81, 83, 209, 224
    on Darwin, 86
    'hopeful monsters', 69, 116–17, 118, 120, 143, 154, 219–20, 222
    cladistics and, 188, 190
single-cell organisms, 7, 8–10, 50–51
    light-sensitive spot, 78–79
skull, comparisons of, 117, 117
slime, 6–7, 15
Smith, Sir John Maynard, 34, 104–9, 116–17, 130, 169, 225
Smith, Professor Sir Grafton Elliot, 182–83
Smithsonian Institution, 20, 21–22
snails, 15
snakes, 74, 82
sociobiology, 51, 52–53, 54
somatic cells, 54–55, 122
Somatic Selection and Adaptive Evolution: On the Inheritance of Acquired Characters, 123
South Africa, 50
South America, x, 21–22, 195, 204–5
space exploration, 135
species stasis, 224
Spirorbis, 137–38

sponges, 7
squirrels, 137–38
  Darwin on, 215–16
Stalin, Joseph, 121, 190
Stanley, Professor Steven M.,
  145
starfish, 15
Stebbins, Leonard, 224
Steele, Ted, 122–23, 124–28,
  passim, 144, 154
stockbreeding, 38–39
streptomycin, 37
Sussex University, 34, 104,
  169
swan, 21–22
  Archaeopteryx and, 21–22
Swann, Mrs. Lucille, 180
Swansea, University of, 129–31
sweet peas
  Mendel and, 32
synthetic theory *see*
  neo-Darwinism
Synthetische Artbildung,
  11–12

tadpole, 118
  axolotl as, 117–18
  eye, 79, *78–79*
tear gland, 66
tektites, 135
Tansmanian wolf, 150–51
Tempo and mode in evolution
  (Simpson), 188–90
Tertiary, 140, 142
Texas University, 116
thermodynamics, 98, 158–59
'Thermodynamics of
  Evolution' (Prigogine),
  162–63
Thom, René, 155–57
Thorpe, Professor William,
  50–51
thyamine, 32–33

tigers, 86–87
Timber wolf, 150–52
Tirpitz, USS, 110
toes, 19–20
  in horse evolution, 16–17, *17*
touraco, 22
trees, fossilized, 101–102
tree of life, 26–29, *26*, *27*
Triassic, 20, 135
trilobite, 14–15, *14*, 45–46,
  95, 136–39
Trinil, 177, 179

ultraviolet light, 35, 46–47,
  48
uniformitarianism, 128–29,
  129–31, 137, 136–39, 140,
  194–95
upright walking, 183, 185–86
United States, 3–4, 38–39,
  42, 45–46, 91–92, 96,
  118, 135
  creationism in, 91–95
  National Academy of
    Sciences, 144–145
  National Science
    Foundation, 142
Ussher, James, 194

Vanderkooi, Dr. Garret,
  91–92
Van Klev, Dr. Harold, 91–92
Velikovsky, Immanuel,
  131–32, 134–35, 138
Vertebrate Paleontology
  (Romer), 13
vertebrates, 20–23, 118–19
  early fish, 6–7, 10–11, 12–14
  tetrapod limb in, 147–148
*Vestiges of the Natural History
  of Creation* (Chambers),
  199–202, 211, 213–14,
  217–18

Victoria, Queen, 197–98, 200–201
Virchow, Rudolf, 178
viruses, 56, 124, 139–145

Waddington, Conrad, 86–87, 130, 150–54, 155–56, 160–61, 175
Wales, University College of, 100–101
'walking-fish' (*Antennarius hispidus*), 11, 19–20
Wallace, Alfred Russel, 211–14, *passim*, 217–18
Warlow, Peter, 133–35, 142
Watson, James, 56
Wedgwood, Emma *see* Darwin, Emma
Wedgwood, Josiah, 203
Weismann, August, 54–55
  barrier, 54–55, 122–23, 127–28
whales, 68–69
  evolution of, 68–70, 77, 79–80
  blubber, 70–71
  eye, 70–71
  mouth, 80
  tail, 70

wheat, 39
  crossbreeding, 38, 121
Wheaton College, (Chicago), 100
Whitcomb, John C., 102
White, A.J. Monty, 101–2
White, Michael, 119–20
Wilberforce, Bishop, 4, 169, 171–72, 218
Williams, George W., 51
William, Mary, 53
Wilson, Allan, 118, 185
Wisconsin, University of, 92
Wistar Institute of Anatomy and Biology symposium at, 64, 86, 151
wolves, 84–85, 149–50
woodcreepers, 21–22
woolly mammoth, 85–86, 121, 132, 134

X-rays, 6, 35, 39, 41, 59

Yale University, 22
Yellowstone Park, 45, 102

Zoology (Darwin), 209–10
Zoonomia (Erasmus Darwin), 196–97

The Study of Man from SIGNET and MENTOR Books

(0451)

☐ **KOSTER: Americans In Search of Their Past by Stuart Struever, Ph.D. and Felicia Antonelli Holton.** A lively narrative describing the ten-year archaeological dig in a cornfield in west central Illinois conducted in 1968 by the author. His find turned out to be a village dating back to 6,500 B.C. that shed light on settlements in prehistoric America. "Should have wide appeal; strongly recommended."—*Library Journal*   (091981—$2.95)

☐ **FATU-HIVA by Thor Heyerdahl.** The fascinating story of Heyerdahl's first Pacific adventure; there were no white inhabitants and no contact with the outside world. "An enormous literary event . . . a work of awesome inspiration . . . simply and beautifully written, engrossing and entertaining."—*Newsday*   (086821—$2.25)

☐ **THE RA EXPEDITIONS by Thor Heyerdahl.** The gripping, day-by-day account of Heyerdahl's epic voyages in papyrus boats across the Atlantic, voyages which created plausible links between the ancient civilization of the Mediterranean and the Americas. Illustrations.
(051211—$1.95)

☐ **THE FIRST AMERICAN: A Story of North American Archaeology by C.W. Ceram.** One of the most popular archaeological writers of our time pieces together the recent discoveries of early American man with "grisly detail, fascinating testimony from diggers and scholars . . ."—*Christian Science Monitor.* Bibliography, Index, and Notes.   (618629—$2.95)

Buy them at your local

bookstore or use coupon

on next page for ordering.

Nature and Man from MENTOR

MENTOR and SIGNET Books of Special Interest

(0451)

Buy them at your local

bookstore or use coupon

on next page for ordering.

MENTOR Books of Special Interest

(0451)

☐ **THE HUMAN BRAIN: Its Capacities and Functions by Isaac Asimov.** A remarkable, clear investigation of how the human brain organizes and controls the total functioning of the individual. Illustrated by Anthony Ravielli. (619013—$2.25)

☐ **THE HUMAN BODY: Its Structure and Operation by Issac Asimov.** A superbly up-to-date and informative study which also includes aids to pronunciation, and derivations of specialized terms. Drawings by Anthony Ravielli. (621166—$2.95)

☐ **THE CHEMICALS OF LIFE by Isaac Asimov.** An investigation of the role of hormones, enzymes, protein and vitamins in the life cycle of the human body. (620372—$1.95)

☐ **THE DOUBLE HELIX by James D. Watson.** A "behind-the-scenes" account of the work that led to the discovery of DNA. "It is a thrilling book from beginning to end—delightful, often funny, vividly observant, full of suspense and mounting tension . . . so directly candid about the brilliant and abrasive personalities and institutions involved . . ."— *New York Times.* Illustrated. (622804—$2.95)*

☐ **INSIDE THE BRAIN by William H. Calvin, Ph.D. and George A. Ojemann, M.D.** In the guise of an operation, we are given a superbly clear, beautifully illustrated guided tour of all that medical science currently knows about the brain—including fascinating new findings. (620526—$2.95)

*Price is $3.50 in Canada
